Praise for *Happiness Genes*:

"Combining recent studies showing the epigenetic effects of changing our beliefs and behaviors with the wisdom of ancient spiritual traditions, *Happiness Genes* is a wide-ranging survey of the science of happiness. It asks the reader provocative questions, like considering how our behaviors might be affecting the genetic expression of our children. Baird and Nadel show you how to inventory your 'happiness assets,' those relationships, possessions and situations that contribute to your happiness, and then treat them like the valuable resource that they are. Wise and practical, the book is peppered with self-reflection exercises, and ends with a 28 day program that guides you in cleaning up everything in your life that stands between you and happiness. To understand the links between what happiness looks like at the molecular level in our cells, the happiness of entire nations, and how you can take lessons from everything in between to raise your own happiness level, this book is an excellent guide."

—Dawson Church,
best-selling author of *The Genie in Your Genes*

Happiness
Genes

INCLUDES THE NATURAL
HAPPINESS 28-DAY PROGRAM

Happiness
Genes

Unlock the Potential
Hidden in Your DNA

JAMES D. BAIRD, PHD
with Laurie Nadel, PhD

Foreword by DR. BRUCE LIPTON, *New York Times* best-selling author of *The Biology of Belief*

New Page Books
A Division of The Career Press, Inc.
Franklin Lakes, N.J.

Disclaimer: *This book and its program are not designed to replace medical or psychiatric treatment for a serious health condition. Please seek professional help if you have questions about your physical or psychological health.*

Happiness Genes
Edited and Typeset by Kara Kumpel
Cover design by Howard Grossman/12E Design
Printed in the U.S.A. by Courier
To order this title, please call toll-free 1-800-CAREER-1 (NJ and Canada: 201-848-0310) to order using VISA or MasterCard, or for further information on books from Career Press.

The Career Press, Inc., 3 Tice Road, PO Box 687, Franklin Lakes, NJ 07417

Library of Congress Cataloging-in-Publication Data
Baird, James D.
    Unlock the positive potential hidden in your DNA / by James D. Baird with Laurie Nadel ;
foreword by Bruce Lipton.
        p. cm.
    Includes bibliographical references and index.
    ISBN 978-1-60163-105-3 – ISBN 978-1-60163-735-2 1.
Happiness. 2. Conduct of life. 3.
Genetic psychology. I. Nadel, Laurie, 1948- II. Title.
    BF575.H27B34 2010
    158—dc22
                                                2009051731

To my wife, Ellie. By her unselfish devotion, she gave me the best memories of my life's journey, and by her example taught me how to love my neighbor.

Acknowledgments

We wish to thank Dr. Wayne Dyer and our literary agent, Lisa Hagan, for believing in this book from the beginning. To Bruce Lipton, our profound thanks for your generous Foreword.

Our appreciation to Dawson Church, Rollins McCraty, Caroline Myss, Gregg Braden, and Steve Bhaerman for their insights into epigenetics. Basha Hayem at Cold Spring Harbor Laboratories allowed a tour of the labs' campus and pointed us toward excellent reference material. To Margaret, Sally, and Deb, who keep Bruce and Dawson organized, our thanks and blessings.

Thanks to Michael Pye and Kirsten Dalley at New Page Books. For each of you who held our hands through the many stages of the editorial process, you have our sincere appreciation.

Finally, to our family and friends for their everlasting patience...we could not have done this without you.

Contents

Foreword
by
Bruce Lipton, PhD

Crisis ignites evolution. The challenges and crises the world faces today are actually signs that change is imminent. We are about to face our evolution.

A paradigm-shattering synthesis of science and society reveals that the planet is in the midst of an incredible evolutionary event...the emergence of a new species, *Humanity*. This evolution is driven by a change in human awareness. We are, each and all, active participants in what will amount to be the greatest adventure in human history: our conscious evolution!

We are living in exciting times, for modern science is in the process of shattering old myths and rewriting the story that has shaped the fate of human civilization. It is a simple fact: Our world changes when our perceptions of life as we know it changes. The old ways of seeing, believing, and reasoning will not help us transform the current conditions, simply because they are the primary reasons for the crises that threaten us today. Changing our understanding of biology and human history offers information and inspiration that will help us navigate these turbulent times.

In order for us to more effectively deal with threatening global challenges and contribute to the evolution of our planet, we must first shed our image of being helpless victims and learn the truth as to who we really are. A good starting place is to first acquire a more accurate account of how life works, for conventional wisdom is being shattered by a revolution in modern science. The new science is toppling the unquestioned pillars of biology, including our perceptions of random evolution, survival of the fittest, and the role of DNA.

the road to empowerment

A fundamental disempowering misperception is evident when we look at ourselves in the mirror and perceive ourselves as single, individual entities. In truth, our body is actually a "community" comprising upward of 50 trillion individual living cells. This number is easy to say, but it is almost unfathomable to comprehend. The total number of cells in a human body is equal to the total number of people on 8,000 Earths!

Virtually every body cell has all of the functions that are present in a human body. For example, almost every cell has its own nervous, digestive, respiratory, musculoskeletal, reproductive, and even immune systems, among others. Each of these cells is, in a true sense, the equivalent of a miniature human being.

Every cell is innately intelligent and can survive on its own when removed from the body. However, when present within the body, each cell foregoes its individuality and behaves as a citizen of a multicellular community. The human body actually reflects the cooperative effort of a *community* of 50 trillion single cells. By definition, a community is an organization of individuals committed to supporting a shared vision. Consequently, whereas every cell is a free-living entity, the body's community accommodates the wishes and intents of its "central voice," a character we perceive as the *mind* and *spirit*.

In a community of cells, the "central voice" represents the equivalent of a government that helps manage the affairs of its citizens. The new science reveals that our minds, replete with thoughts and intentions, profoundly shape our biology, behavior, and genetic activity. An awareness of how our mind influences our genes provides us with an unparalleled opportunity to experience sustainable health and happiness. However, to arrive at this destination, we must first relinquish our old, flawed assumptions about the relationship between mind and body.

a new view of biological control

More than 40 years ago, my research was involved with isolating individual stem cells and placing them into tissue culture dishes. Stem cells are the equivalent of undifferentiated embryonic cells that are scattered throughout your body. Their primary function is to replenish the billions of cells that are lost every day due to normal attrition, as well as repairing damaged or dysfunctional tissues and organs.

The cultured stem cells divide approximately every 10 hours. After two weeks, the stem cell culture dishes contain thousands of cells. The most important point here is that all the cells in a dish are derived from the same parent cell. Consequently, all the cells in a dish are genetically identical. In my experiments, I would split a population of cells into three separate culture dishes, and feed each culture with a different nutrient media containing a unique chemical composition. For cells, the culture medium is the equivalent of their "environment," the same as the air, water, food, and social networks that form our environment.

In one set of dishes the cells formed muscle; in a second set the cells formed bone; and in the third set of cultures, the cells formed fat cells. The big question: "What controls the fate of the cells?" Because all the were genetically identical, the answer was a no-brainer: The *environment* controls the fate of the cells.

This was readily apparent when I would put cells in a less-than-optimal environment; the cells would get sick and the cultures would inevitably die off. To "heal" the cells, we did not use drugs; we simply returned the cells to a supportive environment, and they immediately recovered and the cultures flourished. Simply, the cells adjust their biology to become structural and physiologic complements to their environment.

My research on how the environment controlled the fate of the cells presaged one of today's most important fields of science, *epigenetics*. Epigenetics is the field of study that assesses the mechanisms by which environmental information interfaces the genome and controls genetic activity. Previous to epigenetics, science and the public were preoccupied with a belief emphasizing "genetic control," the belief that genes controlled their own expression. Consequently, we generally speak of the fact that genes "turn on" and genes "turn off," as if genes were self-actualizing and self-regulating.

New insights reveal that genes do not control the fate of cells. They are simply blueprints for proteins, the cell's molecular building blocks. As blueprints, genes do not possess an on or off status. For example, ask an architect working a blueprint a simple question: "Is your blueprint on or off?" The question is absurd.

Genes are blueprints that are activated and controlled by the cell's response to environmental cues, the information contained in the growth medium. Research in epigenetics now recognizes an amazing reality: Changing the chemistry of the growth medium directly alters the readout of each gene blueprint.

In fact, the astonishing finding in epigenetic science is that the environmental information can modify the readout of each gene so as to create more than 30,000 variations from each blueprint. Simply, genes do not represent concrete fates; they are moldable potentials directly influenced by the environment. As the environment changes, so do the biology and fate of the cells.

who's in charge?

How do the findings in stem cell cultures relate to you? As I've described, a human being is not a "single" living entity; we are actually a community of upward of 50 trillion sentient cellular citizens. In truth, we are "skin-covered" Petri dishes containing trillions of cells. The culture medium in our bodies is blood. Consequently, the fate of our body's cells is influenced by the composition of our blood in the same manner that the fate of cultured stem cells are influenced by changing the chemistry of the culture environment.

The big question then amounts to "What controls the chemistry of our blood, which in turn influences the fate of our health and biology?" As I've mentioned, the trillions of cells comprising our bodies are organized into a massive community, within which cells take on specialized functions to support the life of the community. Some cells form specialized heart tissue; other cells form bones, muscles, skin, and blood. The differentiated cells comprising the nervous system are designed to acquire awareness about the world (environment) and use that information to direct the fate and activities of the cellular community.

Specialized nerve receptors, such as eyes, ears, nose, and taste, read environmental information and send signals to the brain. Through the process of "perception," the brain interprets the environmental signals, and in response releases regulatory chemicals into the blood, the body's culture medium. The chemistry derived from the brain circulates throughout the body and controls the behavior and genetic activity of our cells. Consequently, they way we "perceive" our environment controls our health and fate. Most importantly, when we change the way we respond to the environment we change our health and fate.

Under the archaic belief of genetic control we essentially perceived ourselves as victims of our heredity—if cancer or Alzheimer's was in our family lineage, we were led to believe that we should anticipate that we might get stuck with the same fate. However, epigenetic science completely rewrites that limiting belief, for it reveals that through our "mind," we can change the chemistry of our blood, and, in the process, become masters of our fate.

When the mind perceives that the environment is safe and supportive, the cells are preoccupied with the growth and maintenance of the body. In stressful situations, cells forego their normal growth functions and adopt a defensive "protection" posture. The body's energy resources normally used to sustain growth are diverted to systems that provide protection during periods of stress. Simply, growth processes are restricted or suspended in a stressed system. Although our systems can accommodate periods of acute (brief) stress, prolonged or chronic stress is debilitating, for its energy demands interfere with the required maintenance of the body, and, as a consequence, leads to dysfunction and disease.

the value of good driver's education

The principle sources of stress are the misperceptions that have been programmed into the system's "central voice," the *mind*. The mind is like the driver of a vehicle; with good driving skills, a vehicle can be well maintained and provide good performance throughout its life. Bad driving skills generate most of the wrecks that litter the roadside or are stacked in junkyards. If we employ good "driving skills" in managing our behaviors and dealing with our emotions, then we should anticipate a long, happy, and productive life. In contrast, inappropriate behaviors and dysfunctional emotional management, like a bad driver, stresses the cellular "vehicle," interfering with its performance and provoking a breakdown.

Are you a good driver or a bad driver? Before you answer that question, realize that there are two separate minds that create the body's controlling "central voice." The (*self-*) *conscious mind* is the thinking you; it is the creative mind that possesses your wishes, desires, and aspirations. It is also the seat of "positive thinking."

Its supporting partner is the *subconscious mind*, the equivalent of a record-playback device that downloads our life experiences. The subconscious mind is a supercomputer loaded with a database of programmed behaviors. Some programs are derived from genetics; these are our instincts, and they represent *nature*. However, the vast majority of the subconscious programs are acquired through our developmental learning experiences, and they represent *nurture*.

The subconscious mind is *not* a seat of reasoning or creative consciousness; it is strictly a stimulus-response mechanism. When an environmental signal is perceived, the subconscious mind reflexively activates a previously stored behavioral response—no thinking required. The subconscious mind is a programmable autopilot that can navigate our vehicular bodies without the observation or awareness of the pilot—the conscious mind. When the subconscious autopilot is controlling behavior, consciousness is free to dream into the future or review the past.

The dual-mind system's effectiveness is defined by the quality of the programs carried in the subconscious mind. Essentially, the person who taught you to drive molds your driving skills. For example, if you were taught to drive with one foot on the gas and the other on the brake, no matter how many vehicles you owned, each will inevitably express premature brake and engine failure.

Similarly, if our subconscious mind is programmed with inappropriate behavioral responses to life's experiences, then our sub-optimum "driving skills" will contribute to a life of crash-and-burn experiences. For example, cardiovascular disease, the leading cause of death, is directly attributable to behavioral programs that mismanage the body's response to stress.

Are you a good driver or a bad driver? The answer is difficult, for in our conscious creative mind we may consider ourselves as good drivers; however, self-sabotaging or limiting behavioral programs in our subconscious unobservably undermine our efforts. We are generally consciously unaware of our fundamental subconscious perceptions or beliefs about life.

The reason is that the prenatal and neonatal brain is predominately operating in delta and theta EEG frequencies through the first six years of our lives. This low level of brain activity is referred to as the *hypnogogic state*. While in this hypnotic trance, a child does not have to be actively coached by its parents, for children obtain their behavioral programs simply by observing their parents, siblings, peers, and teachers. Did your early developmental experiences provide you with good models of behavior to use in the unfoldment of your own life?

During the first six years of life, a child unconsciously acquires the behavioral repertoire needed to become a functional member of society. In addition, a child's subconscious mind also downloads beliefs relating to self. When a parent tells a young child it is stupid, undeserving, sickly, or any other negative trait, this too is downloaded as a "fact" into the youngster's subconscious mind. These acquired beliefs constitute the "central voice" that subsequently controls the fate of the body's cellular community. Although the conscious mind may hold one's self in high regard, the more powerful unconscious mind may simultaneously engage in self-destructive behavior.

if i am creating my own life... i wouldn't have created this!

The insidious part of the autopilot mechanism is that subconscious behaviors are programmed to engage without the control of, or the observation by, the conscious self. Most importantly, neuroscience now reveals that the subconscious mind, which is an information processor one million times more powerful than the conscious mind, runs our behavior from 95 to 99 percent of the time.

The powerful meaning of this reality is that we only move toward our wishes and desires from 1 to 5 percent of the day. In the remaining time, our lives are controlled by the habit-programs downloaded into the subconscious mind. The most fundamental of these programs were downloaded by observing other people, such as our parents, siblings, teachers, and community. The profound conclusion is that 95 percent or more of our lives is programmed by others!

Additionally, because most of our behaviors are under the control of the subconscious mind, we rarely observe them, much less know that they are engaged. Although your conscious mind perceives you to be a good driver, the unconscious mind that has its hands on the wheel most of the time may be driving you down the road to ruin.

As we become more conscious, we rely less on automated subconscious programs, and also have the ability to rewrite limiting, disempowering beliefs formerly downloaded into the subconscious mind. Through this process we become the masters of our fates rather than the victims of our programs. Conscious awareness can actively transform the character of our lives into ones filled with love, health, and prosperity by its ability to rewrite limiting perceptions (beliefs) and self-sabotaging behaviors.

toward a life of self-empowerment

Through diligent use of our consciousness, we can create lives expressing everything from sublime health to disease. We are no longer victims of forces outside ourselves, nor can we continue to blame such external forces for characteristics and attitudes that we ourselves can change.

In support of our efforts of securing personal empowerment, I truly appreciate the message in Jim Baird and Laurie Nadel's book, *Happiness Genes: Unlock the Positive Potential Within Your DNA*. This valuable work provides the reader with practical ways to apply the insights offered by the new biology. It's the equivalent of a driver's education manual for the human body. This book reveals how you can reprogram your mind and rewrite your genetics to experience a more fulfilling life. It is important to note that we are participants in our own evolution, and therefore we must actively engage in a process to empower our lives.

It's time for new thinking. This book is your wake-up call.

To secure the future we desire, we must empower ourselves with the knowledge of who we truly are: conscious co-creators of our destiny, who are each cells in a new organism called Humanity. We must change our mission from one based on survival of the individual to one that encompasses survival of the species. With an understanding of how our programming shapes our lives and with the knowledge of how we can change that programming, we can rewrite our destiny.

Read this book. Reprogram your thoughts. Learn what makes you truly happy—and make it so.

—Bruce Lipton, PhD
Cell biologist and best-selling author of
The Biology of Belief:
Unleashing the Power of Consciousness,
Matter and Miracles

Introduction

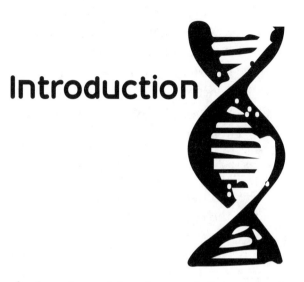

The desire for happiness drives human beings regardless of race, religion, or nationality. Participants in countless surveys usually explain it as their degree of satisfaction or pleasure. But happiness has many meanings. Later in this book, we will define the meaning of natural happiness, for everyone, and how you can accept your gift of happiness. For now, we invite you to examine your meaning through a written exercise.

what is happiness? a preliminary exercise

1. Can you describe how it makes you feel?
2. Do you feel it in your physical body and/or brain?
3. What sorts of activities, beliefs, places, foods, colors, sounds, or people contribute to or reduce your happiness?

4. How do you feel when you do a good deed or an act of compassion?

5. How does climate or the time of day or night influence your being happy—or not?

6. Does your happiness depend on others, or certain circumstances?

7. What is the difference between your feeling of happiness and your pleasures?

The United States Constitution guarantees each individual the right to "life, liberty, and the pursuit of happiness." We have a multi-billion-dollar industry dedicated to persuading you that the secret to happiness lies in money, power, success, or buying the latest gadget, car, or body-enhancement product.

Ask anyone who bought an SUV because he fell in love with a commercial showing it on a mountaintop whether acquiring this vehicle made him or her happy. Perhaps it did briefly...until gas prices rose, making it twice as expensive to fill the tank. Or perhaps until he tried to park the car on a city street or in your average suburban driveway. Similar to millions of his fellow consumers, the SUV owner probably wishes he had turned off the TV instead of sitting glued to the seductive commercials.

The notions of happiness that are fed to us through the media are just that: seductive. They lure us in with false promises that we will be happy when we own a sports car or an iPhone, or have bigger breasts...and when those purchases don't work, the advertising industry flashes new stuff onscreen, urging us to spend more money in the hope that we can purchase happiness.

More than 2,000 years ago, BI (Before Internet) and BTV (Before Television), Gautama Buddha said that "wanting is the source of human suffering." We will never know if he would have succumbed to a six-cylinder eight-passenger conveyance with a built-in GPS navigation system, or if his long, slender fingers would crave the touch-screen experience of an iPhone, but based on the Buddha's teachings, we can speculate that he probably would have turned off the TV as the commercials started.

This leads us to a chicken-egg question: If wanting is the source of human suffering, are we doomed to suffer by wanting happiness?

And, if you don't want happiness, what are the alternatives?

the four assumptions of Western civilization

The second half of this sentence could be considered heretical to the pursuers of materialism...but evidence suggests that the pursuit of happiness according to the following four assumptions invariably leads to disappointment. They are:

1. MORE is better.
2. NEW is better.
3. BIGGER is better.
4. FASTER is better.

When have you bought in to one or more of these assumptions expecting it would bring you happiness? What happened? Would you make the same choice today?

can you buy true happiness?

You need only to scan photographs from all over the world to notice that all humans smile in similar ways. Our mouths turn up, our cheeks puff out slightly, and our eyes glisten. Travel to the Himalayas or the Moroccan desert, and, without knowing a word of the local lingo, you will be able to see, hear, and sense a happy human being.

Without knowing a word of the language, you can communicate to someone whom you have never met before by smiling, laughing, and nodding. Perhaps both of you can point to something funny or lovely on the side of the road. Or you can share delight at seeing something unexpected, such as a horse running down the main street at breakfast time, or a child playing with her younger sister in a field of flowers. Perhaps you were lost and a student who wanted to practice his English walked you to your destination.

In situations such as these, you don't pursue happiness, because it finds you.

when happiness finds you: an exercise

Think of a time when happiness found you.

- Where were you?
- What were you doing?
- Were you looking for happiness at the time?

the search for happiness is universal

Thousands of years ago, the world's wisest teachers gave humanity insight and guidelines for living a happy life. Hieroglyphics in the pharaoh's tombs depict celebrations. The ancient Greeks sought happiness through theater that enraptured the heart with eloquent poetry. In Roman times, too, poets such as Virgil sought to lift the reader's spirit to a level where he was free from care. The Druids of the British Isles and indigenous peoples in the Western hemisphere believed the source of happiness came through the spiritual intelligence of nature. But until modern genetics, the *nature* of happiness remained undiscovered.

the first key

The first key to happiness is understanding that it is inherent in human nature. Each of us can experience it anywhere on the planet, irrespective of culture, language, gender, or race.

Although each human being has a unique way of perceiving emotions and moods, the universal nature of happiness transcends culture, geography, religion, and even chronological time. Who does not feel happy when she performs an act of compassion, such as helping a crippled person across a busy street? An altruistic act fills us with an organic sense of pleasure.

two types of happiness

For thousands of years, the world's wisest teachers have provided insight, tools, and guidance to help students discover happiness. Therefore, it may seem presumptuous to suggest that we in the 21st century need guidance to be able to experience happiness.

The emerging field of positive psychology taught at major universities identifies key components of what we term *common happiness*: pleasures, meaning, and self-improvement. These are subjectively verifiable variables. Although personal experience can validate what is true for an individual, one defect of this model is that the components of happiness are transient and subject to change as each person's life goes through changes. Another, more significant defect is that the components are responsive to external circumstances.

But microbiologists have discovered that there is a biological basis to human happiness, a second type of happiness that is genetic in nature. It is not dependent on external circumstances, but rather is a genetic characteristic. We term this type of happiness *natural happiness*.

Here are some fundamental differences between extrinsic (common) happiness and intrinsic (natural happiness).

Extrinsic/Common Happiness	Intrinsic/Natural Happiness
Pleasure; social in origin	Biological and spiritual in origin
Nurture	Nature
External frame of reference	Internal frame of reference
Happiness comes from the external/material world	Happiness comes from the spiritual/inner world
Action-driven	Faith and insight

Motivated by sense of lack	Motivated by altruistic ethics
"Having"	"Being"
Culturally conditioned	Universal
Behavioral	Genetic/biological
Constitution guarantees pursuit of happiness	Individual's genetic blueprint comes from the Creator
Temporal	Eternal/Timeless
Ordinary	Extraordinary (also non-ordinary)
Material	Spiritual
Conditioned (by expectations)	Inherent

Extrinsic (common) happiness can be pursued by seeking a better job, a bigger home, a faster car, and/or a new relationship. But ultimately, as conditions change, extrinsic happiness leaves us empty and wanting.

Intrinsic happiness occurs naturally. Humans have genes that can be activated by our thoughts, emotional patterns, and behaviors. Mindfulness, meditation, and altruistic actions unlock the DNA code that produces natural happiness.

the second key: you are biologically wired for happiness

For the past 20 years, a mere 3 percent of the mainstream scientific community has devoted itself to the study of consciousness and spirituality. Breakthroughs in physics and biology have been reported and discussed in such books as *The Genie in Your Genes* by Dawson Church, PhD; *The Tao of Physics* by Gary Zukav, PhD; *The Biology of Belief* by Bruce Lipton, PhD; *The Way of the Explorer* by Dr. Edgar Mitchell; and *Sixth Sense: Unlocking Your Ultimate Mind Power* by Laurie Nadel, PhD, with Judy Haims and Robert Stempson.

Dr. Willis Harman, past president of the Institute of Noetic Sciences (*www.ions.org*), called for a new science of consciousness that would establish a scientific model to explain spirituality, intuition, and prayer.

In one of his last interviews before his death in 1994, Nobel scientist Dr. Roger Sperry expressed his belief in a higher intelligence that created and maintained the life-support systems of every human being. One assumption of Western science holds that anything that is real can be reduced to matter in the form of molecules and atomic particles; Dr. Sperry believed that *molecules do not make life-sustaining decisions—consciousness does*. "The molecules of an airplane cannot make it fly," he said, comparing the circuitry to what he called *directive intelligence* within each human being.

a blueprint for happiness

Natural happiness is your biological birthright. It is encoded in your DNA.

Specific genes, such as the DRD4 gene and the VMAT2 gene, are part of the gene pool that provides a blueprint for the intrinsic states of happiness. Further, the existence of DNA strands that are expressed through mindfulness, prayer, and altruistic action provide evidence of a biological basis for spirituality.

It is a common myth that your genes *are* your destiny, by exerting control over your behavior—numerous experiments and studies in the field of behavioral genetics show that interaction with *environment* forms the basis of behavior; your DNA responds to signals from your environment. These signals include beliefs and emotional patterns that "unlock" the positive potential within your DNA. Thus, the New Biology connects the dots among theories of consciousness, cellular communication, and psychology.

your happiness furthers human evolution

You and I are standing on a precipice. At this very moment, we are the first species in the history of evolution that has developed intelligence to a level at which it has the means of making itself extinct. As Bruce Lipton observed in the Foreword, we must begin to participate consciously in the process of evolution. This is a revolutionary idea, but one that is necessary to humanity's survival as a species. We need to evolve together into an empathic community that recognizes that our choices as individuals affect the Whole.

You can begin right now.

By directing your thoughts and emotions to the conscious expression of altruism, mindfulness, and self-transcendence, you create changes in the genetic expression in your DNA. Studies in the new science of epigenetics show that these modifications are saved in the genetic blueprint and transmitted to the next generation of cells!

This means that your happiness is more than a self-directed search for personal joy. As you read further and learn how new patterns of thinking, feeling, and behavior can make you a happier individual, you are also helping humanity to evolve in a peaceful way.

The best part is you don't have to force anything or make anything work. Your genetic blueprint for natural happiness exists within you.

This book aims to further strengthen the bridge between science and spirituality. By building on the work of best-selling New Science authors Gregg Braden and Bruce Lipton, it will inspire and inform you about your spiritual genes and how they work.

It is important to note that simply having spiritual genes does not guarantee behaviors that lead to intrinsic happiness. It is evident from the violence that exists in societies—ancient and modern—that human beings continue to operate from instincts that obstruct the activation of spiritual genes and heart-based belief. Some of those instincts are hard-wired into the biological mechanisms that once helped us survive; others are the result of negative environments and conditioning. To overcome this obstacle, this book provides practical plans that use epigenetic modalities to reduce those negative instincts and environments that tend to block intrinsic happiness.

Despite the suffering and violence inherent in our hectic world, the cause of true, persistent happiness lives within each of us. It is a biological code, imprinted into our DNA, that makes us human and links us to our

Creator. This book will show you, the reader, how to un-lock the code to well-being, contentment, fulfillment, and true happiness.

Author's Note: *If you desire fuller explanations and the history of some of the scientific bases discussed throughout the text, you need only refer to Appendix I.*

Part I:
The Genes
Made Me Do It

chapter 1

Your
Genes
Awaken

*Our relationship to DNA is probably
one of the greatest frontiers which is
being blown open right now.*

—Gregg Braden

Cold Spring Harbor Laboratories (CSHL), a research institute for molecular biology and genetics, spreads across 130 acres of well-groomed waterfront property on the prestigious northern shore of Long Island, New York. An inlet to Long Island Sound, a ribbon of quiet water, punctuates the property. The town of Cold Spring Harbor for which the institute is named is a former whaling port that Native Americans, its original residents, named Wawapex, meaning "Good Little Water Place."

Since 1890, when a group of wealthy local investors founded the institute, the laboratory at Cold Spring Harbor has produced seven Nobel Prize winners. Some 85 Nobel laureates have worked here at some point during their illustrious careers.[1] It was here, in 1953, that James Watson gave his first public lecture on the double-helix structure of the DNA molecule. Nine years later, he and colleague Francis Crick won the Nobel Prize in Physiology of Medicine for identifying the structure of DNA.

Today, some 120 years after the laboratory was founded, genetics researchers from all across the globe come to Cold Spring Harbor Laboratories for conferences and symposia on multisyllabic topics pertaining to microscopic phenomena that include genes and cancer research, neuroscience, and plant biology. Bioinformatics, a new field that examines biological data and statistics using state-of-the-art computer programs, is one emerging field of study. Another is epigenetics, which we will get to in just a moment.

a historical perspective

If you have ever visited the old city of Philadelphia, or colonial Williamsburg, you can get a sense, during a walking tour, of what occurred in the past. In a similar fashion, the history of science was made in Cold Spring Harbor Laboratories. Nobel Prize–winners walked these paths on their way to lunch, thinking deeply about their scientific projects. Even a non-scientist taking a walking tour of the facilities cannot quite escape the energetic presence of great scientific minds, past and present.

With the official opening of the Carnegie lab building in 1905, Cold Spring Harbor Laboratories became one of the first genetics research institutions in the world. The Carnegie building now houses historical documents, including an outline that helped to launch the Human

Genome Project, an international program that mapped all the genes in the human body.

Nearby, the Jones building is the country's oldest biology lab in continuous use. A look inside reveals an original floor-to-ceiling fireplace, and a few scientists studying rat neurons under microscopes. The open space hums with activity, although the giant steel pods encased in glass are designed to result in a vibration-free facility. Jones has won several architecture awards for "adaptive re-use" of an old building for contemporary needs.

After peeking inside one or two labs, you may be wondering what a campus full of science geniuses would look like. Picture, if you will, some 400 versions of Dr. Gregory House: brilliant, analytical, and relentless in their commitment to solving complex problems.

To an observer, interaction among colleagues is not what you would expect. Whether engrossed in their lab work or walking to the dining hall, scientists on campus rarely make eye contact with each other, much less with a visitor on tour. Their eye movements dart upward and sideways, indicating that they are focusing more attention on their ideas than on the person who is speaking. Although their conversations are subdued in volume, they are animated in pace and intensity. From time to time, a scientist will flash an inscrutable smile as his or her eyes light up in a moment of "aha!" just like House when he has figured out what's wrong with his patient. But there are 400 brilliant minds on the premises, cogitating like mad to come up with genetic solutions to diseases such as cancer, diabetes, and Alzheimer's.

At the turn of the 19th century, genetics research was primarily focused on eugenics, a field of research that has since fallen out of favor. *Eu-* is the Latin prefix for "good," and *genics* refers to shared traits. Eugenics was believed to work by "encouraging reproduction by persons presumed to have inheritable desirable traits," and discouraging reproduction by those who had inheritable traits that were not so desirable.[2] (Carried to extremes, Hitler's so-called master race was predicated on the eugenics model.)

as any *CSI* fan knows...

DNA is to today's TV crime shows what Desdemona's handkerchief was to Shakespeare's Othello. Desdemona's dropping her handkerchief became the key to understanding the events that unfolded; likewise, a perp's DNA in this week's episode of *Law and Order* or *CSI,* serves as a biological tag that helps pathologists to identify crime victims and perpetrators. Like a fingerprint, every human's biological tag is unique. According to *DNA.gov,* "DNA, or deoxyribonucleic acid, is the fundamental building block for an individual's entire genetic makeup. It is a component of virtually every cell in the human body."[3]

DNA is a code. Massive groups of DNA make up your genes, and your genes are like instructions that can activate a sequence of code for heritable traits embedded in a microscopic strand of DNA. The DNA contains the words and structure of the language that is used for a particular set of instructions. Your genes can unlock the code within your DNA, not the other way around. A few paragraphs from now, We will give you an example of just how this works.

Your 3 billion DNA molecules are encased in protein sheaths. Certain genes, called *regulatory genes*, contain the instructions to wrap or unwrap the protein sheath. This, in turn, can activate or stop the DNA code from going into effect. In searching for genetic-based cures for certain diseases, researchers are interested in finding out which genes can "lock" or "freeze" the activation of DNA codes for those illnesses. If you can switch off those genes, the DNA code will not be able to execute that particular disease's program. Scientists believe that, when they discover the gene mechanisms that control the disease's DNA, they may be able to develop or design molecules that will be able to turn off those genes and halt the onset of a specific disease.

A molecule that switches off a regulatory gene is creating a change in the biochemical environment. Temperature, toxins, and pollutants can also produce environmental changes that alter the activation of certain genes and, subsequently, specific strands of DNA as well.

what you learned in high school about DNA is outdated

If you have ever undergone an EEG test, you would have seen your mind's output as a series of waves on paper: Your brain's output is measured in electromagnetic waves. Your variable heart rate produces a measurable *magnetic* field as well, and these *magnetic* signals are another type of environmental trigger that can affect how your genes behave. Studies now prove that you can deliberately and consciously change your baseline emotional level and thoughts so as to change your body's measurable *magnetic* output—in a way that can affect whether or not your genes go into action to instruct strands of DNA that contain certain codes. Every cell in your body contains DNA, and the structure of your DNA does not change. However, the genes made up of masses of DNA can be altered by the external physical environment, and by your mind and your emotions, your internal environment.

That's not what I learned in high school/college/graduate school/on TV! you may well be thinking.

No, it is not. Those concepts are outdated. Remember: People used to believe that the Earth revolved around the sun and that evil spirits, not bacteria, caused disease. The spirit of science is about searching for truth and discarding old beliefs in the light of new facts that prove those beliefs to be no longer valid.

As you are about to see, new discoveries in the lab challenge what you learned in your high-school biology class. There are millions of DNA molecules in your body serving an uncountable myriad purposes. Some of these have been studied according to rigorous standards of science and found to be affected by your internal environment.

From a mind/body perspective, this brings good news and bad news: The old model let you blame your circumstances on your genes, as in "It's not my fault. I have lousy genes." The bad news is that you can no longer make excuses for your unhappiness by blaming your DNA, but the good news is that you can discover new ways of directing the flow of your emotions and thoughts to improve your quality of life while creating powerful biophysical changes at the genetic level.

Are you ready to see how it can work?

a landmark study on switching off cancer genes

At the University of California, San Francisco, Dr. Dean Ornish and his research team studied a group of men with prostate cancer who had chosen not to have surgery, radiation, or chemotherapy. For three months, the subjects walked for half an hour every day, ate vegetables, soy, and whole grains, and practiced meditation for stress reduction. The results of the study, published in *Proceedings of the National Academy of Sciences* in June 2008, showed that, in addition to lowering their blood pressure and losing weight, the men's genes for prostate and breast cancer had shut down. He noted, "It's an exciting finding because so often people say, 'Oh, it's all in my genes, what can I do?' Well, it turns out you may be able to do a lot."[4]

welcome to the field of epigenetics

The study of gene interactions is called *epigenetics*. *Epi-* is the Greek prefix meaning "above" or "higher than," as in *epidermis*, the scientific term for your skin. Conrad Waddington gets credit for coming up with the term *epigenetics* to describe a model of how genes interact with their surroundings. But the ancient Greek philosopher Aristotle (384–322 BC) referred to *epigenesis* as the development of "an individual organic form from the unformed."[5] Thomas Jenuwein, an Austrian scientist, likens the difference between traditional genetics and epigenetics to "the difference between writing and reading a book. Once the book is written, the text [DNA] will be the same in all the copies.... Epigenetics would result in different read-outs."[6]

One of the hottest fields of science today, epigenetics explores how your genes behave when their environment changes. As the Ornish study shows, a regulatory gene for cancer can be switched on and off by environmental modifications. "DNA is just tape carrying information, and a tape is no good without a player. *Epigenetics* is about the tape player," observes British scientist Bryan Turner.[7]

The majority of epigenetics research looks at chemical interactions that alter how your genes behave. With funding from pharmaceutical companies, scientists are being paid to look for molecular magic bullets. But, as you will see shortly, psychological shifts can modify gene behavior as well.

Changing a gene's environment can trigger a type of molecule called a *methyl group*. The *methyl group* can stop or start the gene from activating its set of instructions. According to genetics researchers at Duke University, methylation can be compared to "putting gum on a light switch...the switch isn't broken, but the gum blocks its function."[8]

In his groundbreaking book *The Biology of Belief*, Bruce Lipton describes how he discovered that the membrane of a human cell contains a transmitter and a receiver for messages in the form of electromagnetic signals from other cells. Electromagnetic signals can communicate instructions to such molecules as a methyl group. In turn, a methyl group can act upon a gene so as to interrupt its normal behavior. "DNA is like a program in your computer. It contains instructions," says Dr. Lipton. "It is not—repeat—*not* the hardware."

when rats make bad mothers

From a layperson's perspective, genetics and epigenetics raise a basic question: nature or nurture?

"We can no longer argue whether genes or environment has a greater impact on our health and development, because both are inextricably linked," said Randy Jirtle, PhD, a genetics researcher in Duke's Department of Radiation Oncology. "Each nutrient, each interaction, each experience can manifest itself through biochemical changes that ultimately dictate gene expression, whether at birth or 40 years down the road."[9]

At an epigenetics conference sponsored by Duke University in October 2005, Moshe Szyf, PhD, professor of pharmacology and therapeutics at McGill University in Montreal, and his colleague, Michael Meaney, studied maternal grooming behaviors in rats. Baby rats that were not licked by their mothers produced more stress hormones as adults than baby rats who received proper nurturing for their species. Dr. Szyf found that *the gene that releases the protein responsible for regulating the production of hormones linked to stress had less methylation!* In other words, the mother rat's lack of proper nurturing behavior limited her offspring's genetic ability to produce enough proteins to lower the offspring's stress hormone levels. "We're showing that it's the maternal behavior that counts, not just the genetic baggage," he said. "Behavior can clearly affect the chemistry of DNA."[10]

a behavioral epigenetic link between child abuse and suicide

Scientists may have left the Duke University epigenetics conference wondering whether epigenetic factors would have similar effects on humans. In fact, after examining the brains of 12 suicide victims who were abused as children, Dr. Meaney proved, four years later, that abuse during one's childhood can damage an individual's genetic ability to produce a normal flow of stress hormones. Autopsies of the victims' brains, plus brains of other suicide victims who had *not* reported having been abused as children, and brains of people who died of natural causes, showed significant differences. Researchers concluded that child abuse had a destructive effect on a genetically controlled stress response mechanism called the *hypothalamic-pituitary-adrenal axis* (HPA). Normally, HPA activity is regulated by a gene called a *glucicocorticoid receptor gene*.[11] Dr. Meaney's 2009 study found that this gene had produced smaller amounts of the protein that regulates HPA activity in these 12 suicide victims. They concluded that childhood abuse led to this gene's stunted ability to produce the correct amount of a protein that would have stabilized the production of HPA. In other words, this gene should have functioned as a surge protector, but in the 12 suicide victims who had suffered child abuse, the surge protector was not working properly. In comparison, the brains of suicide victims who had *not* suffered abuse in their childhood showed normal amounts of the genetically regulated protein that inhibits HPA activity. So did the brains of people who had died from causes other than suicide who had also not reported having been abused as children. Dr. Meaney's research team concluded that violent and neglectful parental behavior had an impact on "the epigenetic regulation of glucocorticoid receptor expression."[12]

One impressive implication of this study is the need to ensure that parents receive effective treatment for mood disorders and substance abuse that are often linked to abusive behaviors. Parents who were themselves abused as children are especially at risk.

The Human Genome Project

Your *genome* consists of all of your genes—approximately 30,000 genes that form 3 billion pairs of DNA molecules.

When it was officially launched on October 1, 1990, the Human Genome Research Institute (NHGRI), the Department of Energy Office of Biological and Environmental Research, and the partner organizations in the International Human Genome Sequencing Consortium set a 15-year deadline for mapping the human genome. But the project came in two years ahead of schedule on April 14, 2003, and well under budget. Research institutes, including Cold Spring Harbor Laboratories and universities in the United States, China, France, Germany, Japan, and the United Kingdom participated in the global effort to identify and map all the genes in a human being.

With the mapping phase of the project completed, the Human Genome Project now provides researchers with

new information and data for bioinformatics specialists to analyze. As genetics researchers continue to study
the genome, new information is posted on the Human
Genome Project's Website every 24 hours.[13] Because
the implications for health, preventing and curing disease, and cloning raise social, legal, and ethical issues, the
Human Genome Project maintains links to the latest dis-
cussions on these complex themes.[14]

The complexity of the research being done by the
HGP meant that DNA donors to the project needed to
provide more than that swab of saliva your average TV
crime show needs to identify someone. For the genome-
mapping program, the DNA came from blood that was
donated by thousands of anonymous volunteers recruited
through public advertisements posted near the research
laboratories where the DNA "libraries" were being pre-
pared. More than five times as many individuals donated blood than was actually used. The secrecy surrounding this project was so air-tight that the Human Genome
Project maintains that "not even the volunteers would
know whether their sample was used."[15]

your genome, your health

The Human Genome Project is continuously producing new information that may eventually lead to individualized programs for improved health and preventive
medicine. For example, if you have a gene for an illness
that could appear later in life, you will be able to decide if
you want to use a self-care program based on epigenetics
research, or perhaps a drug or an herbal remedy, that can
deactivate, or "turn off" the regulatory genes that contain

the instructions for your DNA to manifest a disease such as diabetes or cancer.

A genetic variation mapping project known as the National Human Genome Research Institute's "HapMap" continues the search for genetic clues to such diseases as asthma, Alzheimer's, and coronary heart disease. Researchers are also using HapMap to search for environmental influences that can alter the gene's behavior mechanisms.

Dawson Church, PhD, author of *The Genie in Your Genes*, leads research teams looking for genetic changes that occur when a group of subjects' emotional states change. He and his team have found 48 genes that predict post-traumatic stress disorder (PTSD). Using an acupressure technique called Emotional Freedom Technique with subjects who suffer from PTSD, researchers are getting ready to study the activity of those 48 genes before and after each intervention. The Emotional Freedom Technique is considered "energy psychology" because it includes tapping certain places on the body that correspond to acupuncture points that, in turn, release energy in order to relieve such PTSD symptoms as flashbacks, avoidance, and hyper-vigilance.

"Experimentally we know that energy psychology is a potent epigenetic intervention for anxiety, depression, physical pain, and PTSD," says Dr. Church. He is, however, quick to caution: "Behavioral epigenetics serves as a curiosity at this point. Most of the work is being done following the old drug model: Find the gene that can be silenced with a pharmaceutical magic bullet even though there are long-term effects that can be damaging to your health."[16] But as the Ornish study shows, you can produce positive genetic changes with diet, exercise, and meditation—interventions that do not produce profits for drug companies.

Interventions such as meditation, walking, a healthy diet, acupressure, kinesiology, hypnosis, and neurolinguistic programming may offer safe, natural, and effective alternatives that use your own physiological epigenetic mechanisms to boost your brain's production of serotonin, the neurotransmitter that can elevate your baseline mood state to a new, sustainable level of happiness and well-being.

looking ahead

The science of epigenetics studies how your genes behave, and has clinically shown that modifications to the physical (external) and mental (internal) environment—heat, pollution, chemicals, and an altered mindset—can alter specific DNA molecules.

Considered one of the hottest new sciences, epigenetics represents the future of medicine and individualized healthcare. Not only will it be possible to use drugs or herbal remedies to switch off genes that, in turn, are programmed to activate the DNA for certain diseases and deficits, but epigenetics will furthermore give you choices that are safe, natural, and effective—with no damaging or life-threatening side-effects.

Are you ready to explore your new choices?

questions for the reader

To gain a fuller understanding of the concepts and information in each chapter, you may wish to reflect on these questions and write your thoughts in a journal. The Socratic method of thinking in open-ended questions can also serve as a strong methodology for classrooms, workshops, and book groups.

- What do you know about your DNA?
- Do you believe that you can develop—or neglect—a genetic trait, such as musical or athletic ability?
- Do you believe that all of your DNA cannot be changed?
- How do you feel about the following statement: "I am a victim of my genes"?

chapter 2

Epigenetics, Evolution, and You

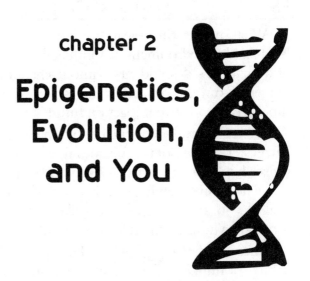

*This proclivity toward violence and greed
is psychospiritual. As each individual turns
inward and undergoes deep transformation,
the species evolves. Hopefully in time.*

—Stanislav Grof, MD

As you begin to consider that the biological instructions encoded in your genes are not always destined to activate, you might wonder how your thoughts, emotions, and actions could create a potential ripple effect on future generations.

Although you cannot control the contents of your individual genome, there are various choices you can make that will enhance your health and improve your mood. As demonstrated in Dr. Ornish's study, a healthy diet, half an hour of walking every day, and basic meditation can actually turn off hundreds of cancer genes. Who would not be happy about that?

What might this mean for you and your family? Because the environment can influence the development of a genome's descendants without changing the DNA sequence, can the positive changes we make in our lifetimes have a positive impact on future generations?[1]

A recent study conducted by the Mayo Clinic in Rochester, Minnesota, concludes that, when your health is excellent, you are twice as likely to be happy as someone who is in "merely good health." If you are in poor health, you are 70 percent less likely to be happy than someone who is in "merely good health."[2]

These studies suggest that when you cultivate "happy habits" of thinking, feeling, and behavior, you can generate the kind of epigenetic changes that can influence the course of evolution!

Instead of acting out of those darker instincts, you can take advantage of innumerable options of spiritual expression, cognitive tools, altruistic actions, mindfulness, prayer, nutrition, and physical exercise that will serve to raise your emotional baseline—not just now, but potentially for generations to come. Further, the Mayo Clinic study finds that those activities you choose intentionally, mindfully, and proactively are likely to improve your baseline happiness for the long haul.[3]

In addition to helping scientists discover the molecular keys that can turn off genes for specific diseases, the epigenetic frontier is wide open for researchers who are looking for other-than-chemical interventions to alter the environment controlling certain genes. "The epigenome's malleability highlights the power we can have over the health of ourselves and our children," said Dr. Jirtle.[4]

If you recall the rat study we mentioned in Chapter 1, the rats that were neglected by their mothers grew up with higher stress levels and anxiety than baby rats that received the right amount of maternal affection in the form of licking. This shows that, not only does an absence of affection during childhood impact the adult rat's genetic ability to regulate the flow of stress hormones, but it also shows how a baby rat's epigenome functions by activating the regulatory genes in question. In other words, affectionate maternal behavior makes the baby rat grow up calmer, even during stressful events.[5] As you can see, the epigenome's responsiveness to interpersonal relationships provides further support to those numerous studies that show that solid friendships and social support are essential for your well-being and happiness. Even the genes say it's so!

an evolutionary crossroads

Despite our scientific knowledge and technological advancements, when driven by territorial instincts, human behavior harkens back to our Stone Age ancestry. The drive to survive has traditionally led to an "us versus them" mindset that, in turn, could become violent and destructive, thus obliterating any potential for personal happiness.

In one of his last interviews, Nobel Prize–winning scientist Roger Sperry, MD, told journalist Laurie Nadel, "Sometimes I think we are evolving into a less intelligent species. We have become the first species in the history of evolution that has used its intelligence to invent the means to make ourselves extinct. I find that very depressing." Dr. Sperry's concerns have been echoed by any number of prominent scientists whom Laurie has interviewed, including Jonas Salk, former astronaut Edgar Mitchell, and biologist Bruce Lipton.

In *The Anatomy of Reality: The Merging of Intuition and Reason*, Jonas Salk, the doctor who invented the Salk polio vaccine, wrote, "The evolution of the human mind... depends upon the evolution of intuition and reason."[6] If humanity is to succeed in continuing to evolve, we need to value empathy and interconnectedness as much as logic and reason. Until we begin to perceive that our individual beliefs, emotions, and actions affect other human beings, we are in danger of reasoning ourselves to extinction. Dr. Salk believed that humanity has the potential for conscious evolution, rather than simply evolving as a process of physical survival. According to what he called his "biophilosophy," evolution needed to become conscious and mindful as well as biological. Before his death, he called on scientists and researchers in the humanities to work together "to make the decisions and choices that Nature has made until now...for the greatest value to human life and society as a whole."[7]

from outer space to conscious evolution

Former astronaut Edgar Mitchell had an epiphany in space when he recognized the presence of a Higher Intelligence behind the order of the universe. He left NASA in 1973 and founded the Institute for Noetic Sciences (*www.noetic.org*), a research organization that studies consciousness, intuition, and such phenomena as mind-body healing. In *The Way of the Explorer*, he presents his ideas on how our consciousness can impact human evolution. Taking those ideas a step further, he says, "Our evolution is a learning process. When we begin to practice meditation, for example, we open to the reality of the transcendence experience described by yogis, prophets, and mystics of all religions. We experience ourselves as interconnected to others through God. This, in turn, causes a shift in behavior. We come to abhor violence."[8]

do you have "victim chemistry"?

Scientist and *New York Times* best-selling author Gregg Braden agrees. "To me, this makes tremendous sense: We come into a world with the ability to somehow participate in this world. It makes no sense that we would come into a world helpless, powerless, and just be the victim of whatever happens to come along in the world or in life," he says.[9] Although epigenetics points to the importance of DNA, the code still needs to receive instructions that tell it how to perform in the human body. "Those instructions come from the environment around us—our physical environment, as well as our emotional environment of thought, feeling, belief, and emotion. These are now directly linked to allow the genes within our bodies their fullest expression, or preventing that expression," Braden said, noting further, "When we believe that we are powerless victims, that victim chemistry is released into our bodies. We've got a low-immune response and our body may not be as strong and resilient. After all, that's the belief that creates the electromagnetic signal that instructs the immune system's DNA."[10]

The change in human evolution that needs to take place will come from people changing how they feel. "When we feel safe in our world, and we learn to love rather than fear, these are literal codes that our hearts send to our brains," Braden says. The brain responds to those signals by giving the body what he calls "life-affirming chemistry, life-affirming hormones, and a strong immune system." Although our beliefs play a powerful role in activating the genes that trigger our body's optimum response, Braden observed, "What science is showing and what the ancient traditions have always said is that this is how we participate in building the kind of world we would all like to live in; not by controlling or manipulating or imposing our will. The beauty is, we don't have to know any of that."[11]

Recent epigenetics studies show that your life experiences and your environment can impact the behavior of certain genes. As you have seen, parents who are emotionally unstable or neglectful can unknowingly turn off their children's genes that modulate the flow of stress hormones. Children who are biologically impaired in this way are hardwired for high levels of chronic anxiety, posttraumatic stress disorder, and emotional instability. "It is well known that parental behavior affects children. My paper suggests a way that the parent's psychology before conception can actually affect the child's genes," said Dr. Alberto Halabe Bucay of Mexico. Dr. Halabe published a paper in *Bioscience Hypotheses*, a scientific journal that features intriguing hypotheses rather than scientific studies that have been through a peer review process.[12]

As you cultivate habits of thinking, emotion, and action that make you happy, you reinforce certain personality traits that can be inherited by your children. "Although happiness is subject to a wide range of external influences, we have found that there is a heritable component of happiness which can be explained by the genetic architecture of personality," said Scottish psychologist Alexander Weiss.[13] In other words, some traits, such as a sunny disposition, may be due, in part, to your parents' genes as well as what happened during your childhood. Furthermore, your parents' genes may have been altered in response to how life treated them, thereby passing on some of their impaired genes to you.

The great news is that you can use this information to build happiness into your life on a consistent basis. As you do, not only will your life become more enjoyable, but your health will improve and your own happiness genes will be activated, expressed, and passed on to your children and hopefully their children's children as well.

your heartfelt emotions and your DNA

Since 1990, the Institute of HeartMath in Boulder Creek, California, has been conducting physiological research into the merging of heart and mind. Specifically, HeartMath has developed technology and protocols to measure what Rollin McCraty, PhD, calls "the intelligence of the heart." A founder of the Institute of HeartMath, Dr. McCraty's studies have produced scientific information that proves what centuries of literature, music, and art have expressed: "The heart is the gateway to the spirit."[14]

Using a device called an emWave monitor, scientists have developed a model for what they call "a physiological state of coherence associated with such positive emotions as appreciation, caring, and compassion." This new form of biofeedback technology measures a person's heart rate variability (HRV). "If you look at the pattern that the heart beats out, it reflects our current emotional state. If it is less than it is supposed to be, it can be an indicator of depression or illness. When we are angry, the HRV reflects a dys-synchronicity," says McCraty.[15]

When you are happy, your heart rate variability pattern is smooth and sine wave–like. When the messages the heart sends to the brain are coherent—high and even—the brain interprets that as a good feeling.

The magnetic field generated by the heart is far greater than the magnetism generated by brainwaves. This magnetic field has encoded information that is relevant to your emotional state. It functions like a carrier wave in a radio. McCraty says, "All of our DNA in all our cells is exposed to that magnetic field from the heart!"[16]

Using the emWave heart-rhythm feedback technology developed by HeartMath, their researchers are now able to measure one person's heartbeat and another person's brainwave when they are up to 5 feet apart. They can measure what happens when those heart- and brainwaves synchronize, creating empathy.

Studies of individuals using a personal emWave device to achieve this state of coherence have reported such additional benefits as optimal cognitive functions, sensory motor integration, a stronger mind/brain connection, and sharper intuition. "These tools and techniques increase our ability to self-regulate from a more intelligent baseline perspective," says McCraty, who describes the process as "a practical tool to actually do the things people say they should do but rarely know how to do—the, A-B-C of how to do that."[17]

Lest you think that *coherence* is just another word for meditation, McCraty observes that most meditation states do not lead to states of coherence, which he further defines as "being here, in charge, and more proactive." When you are fully in a state of kindness and compassion there is a measurable physiological state that is reflected in the variable heart rhythm. "It's not about thought," he says. "It's about emotion."[18]

A series of recent laboratory experiments conducted by McRaty and his associates have come up with evidence that DNA can change as the result of a subject's focused intention and positive emotional state. Using an ultraviolet spectrophotometer to measure a specific wavelength that DNA absorbs, the scientists used heat to partially unwind a strand of DNA that was used as a target. Subjects included healers, university students, and random people from the street who were invited to take part in the experiment. Researchers discovered that two important variables are necessary for reliable results. In order to intentionally cause DNA to wind or unwind, it is necessary for an individual to be in a state of what McRaty calls "whole system coherence" in which the heart rhythm reflects a measurable state of heart coherence and holds the intentional thought of wrapping or unwrapping the DNA. "You could be in a highly coherent state without the intention or you could have the intention in your mind but not be in a state of heart coherence, and nothing would happen to the DNA," says McRaty.[19]

As that old song goes, "You gotta have heart!"

can you change the course of evolution?

It's a leap of the mind to think that because a handful of scientific experiments have shown a direct cause and effect between positive emotional states and genetic changes that you can change the course of human evolution by becoming a happier person.

But really, what do you have to lose?

Anxiety? Depression? Insecurity? Low self-esteem? Addictions? Destructive patterns of behavior, including, but not limited to, violence toward your children?

And what do you have to gain?

A positive outlook on life? A more upbeat emotional baseline? Consistent optimism? Resilience in the face of adversity? The ability to perceive opportunities instead of feeling overwhelmed by obstacles? Healthier habits? Improved immune-system functioning? Better health? A happier family? Finding pleasure in your everyday life? Knowing that your choices can turn certain genes off and on, leading to a healthier future for yourself and future generations?

taking the first step

Have you ever asked yourself, "How *good* can it get?"[20]

Why not?

What happens—in your heart, in your mind, and in your entire being—when you ask yourself, "How *good* can it get?"

What would happen if every time you brushed your teeth or washed your face, you looked at your reflection in the mirror and asked, "How good can it get?"

What if you taught your children to do that every day and every night?

We can't predict the future, but we can't help thinking that you and your family will be happier...and so will your DNA.

conclusion

Yes, these studies in epigenetics represent the first baby steps in this new science. But the findings have implications that you can adopt to make positive changes in your own life.

questions for the reader

- How can this research help you to become a happier person?
- What are your reactions to the implication that your happiness can have a positive impact on future generations?
- Do you agree or disagree with the concept of conscious evolution?
- Do you think that the human species is evolving into a more or a less intelligent species?
- What would you like to take away from this chapter for use in your own life?

chapter 3
The Power of Our Spiritual Genes

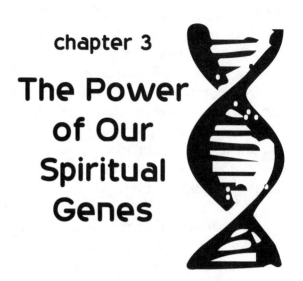

*States of self-transcendence are essential
if the human species is to survive.*

—Edgar Mitchell

Before new scientists began finding proof that we are not victims of our genes, language was the only method for codifying a human's subjective experience. Such terms as *love your neighbor*, *faith*, *hope*, *soul*, *spirit*, *heart*, and *essence* were used to ascribe meaning to spiritual states of being that flowed into such heart-based states as happiness, joy, ecstasy, inner peace, and divine purpose.

Recent scientific discoveries in the fields of microbiology, cellular behavior, and genetics take nothing away from subjective experiences; rather, these findings provide evidence that spiritual states linked to happiness have a biogenetic foundation. It is your God-given biological birthright to develop habits of happiness. These habits empower the expression of spiritual genes that form this biological blueprint for fulfillment and joy.

Furthermore, these discoveries add to the extant knowledge of human nature. They show that our body, mind, and spirit are truly interconnected. Your spiritual DNA is waiting to be awakened so that your life— and the lives of others—can be enhanced as your soul awakens to its true purpose: honoring your connection to the Creator. Throughout recorded time, in every culture, both current and extinct, humans have sought connection with this invisible power that is greater than any one human or human collective. The Creator is known by many names: God, the One, the Source, the Universe, the Higher Power, and Spirit (just to name a few).

We now know that true, intrinsic happiness comes from expressing a spiritual instinct that has a biological basis. To be human is to seek God.

Our map lies within.

a biological treasure map

In her brilliant book *Molecules of Emotion*, Candace Pert proved that emotions are stored as molecules in the limbic system of your brain. The limbic system is also known as "the brain of connection" and "the emotional brain."

The most powerful emotional experiences are those that bring joy, inspiration, and the kind of love that makes suffering bearable. These emotional experiences are the result of choices and behaviors that result in our feeling happy. When we look at happiness through a spiritual filter, we realize that it does not mean the absence of pain or heartache. Sitting with a sick or injured child, every parent gets to know the profound joy that bubbles over when a son or daughter begins to heal. This is a simple example of how we can be flooded with happiness that becomes more intense as we contrast it with previous suffering.

Experiences such as this go into the chemical archives of the limbic system. Each time you experience true happiness, the stored emotions are activated as you are flooded with even deeper joy than you remembered. Your spiritual genes are, in a sense, your biological treasure map to joy.

You can become a happier person by cultivating beliefs and behaviors that reinforce those molecules of powerful and transcendant emotional states so that your baseline emotional state gets uplifted. In other words, if you wake up every morning feeling about a 4 or a 5 on a scale of happiness where 10 is your maximum and 0 means you are not feeling anything resembling happiness, you can cultivate new habits of thinking and actions that will elevate your baseline emotional state to a 6, 7, or even an 8. We would like to say that with practice, you can elevate your baseline emotional state to a 9 or a 10.

Try this experiment:

- When you awaken tomorrow morning, check in with your mood.
- Rate your happiness on a scale of 0 to 10. Write it down, with the date.
- Keep a log for seven to 10 days to determine your baseline state.
- As you read further and try the exercises in the last chapter, keep a log of your waking baseline state for another week to 10 days.

You may well be pleasantly surprised by what you find.

questions for the reader

- What words do you use to describe happiness?
- "To be human is to seek God." Do you agree or disagree? Why or why not?
- What is your baseline emotional state?

Part II:

Love That

Feeling

chapter 4

Your Ancestors' Spiritual Instinct

In the ancient world people believed that it was only by participating in this divine life that they would become truly human.

—Karen Armstrong, *A History of God*

Thousands of years ago, one of your Stone Age ancestors scratched a star, moon, sun, and a cross into the wall of his cave. It was so easy, even a caveman could do it!

A modern-day analysis of those crude squiggles show, that, despite what you may have learned in school, there was more going on in a Neanderthal's brain than merely hunting for food, clothing, and shelter. Nor should you forget there was yet another instinct that drove your ancestors: fornication. Without the drive to procreate, who knows? You might not be here in your present state of human evolution. You might not even be here at all.

Until quite recently, those basic instincts were all that could be attributed to humans living some 70,000 years ago. But in the past decade, a quiet revolution has been shaking up the assumptions upon which scientists interpreted evidence about the lives of your Stone Age forebears. Not only were they hunting, protecting, and procreating with grunts and guttural sounds, but the first humans left artifacts and crude markings that point to yet another biological instinct, one that has profound implications for you, here in the 21st century.

archetypes and the spiritual instinct

Archetypal symbols, such as the moon, stars, sun, or a basic cross, provide visual evidence that even before man had language, his genetic code contained a biological instinct for spirituality. Although it was not directly connected to his physical needs for food, clothing, shelter, and sex, your ancestor instinctively reached for a connection to an invisible, mysterious force greater than any human being.

In 2006, archeologists discovered the earliest evidence of man's spiritual instinct. A python-shaped carving, consisting of 300 indentations etched to resemble the snake's scales, body, and head, had been carved into the rock wall of a cave in the African nation of Botswana some 70,000 years ago. "We believe this is the earliest archaeological proof of religion," Sheila Coulson, a Canadian expert in Stone Age tools told Reuters News Service.[1]

The python symbol is considered powerful and sacred to the San people of northwestern Botswana, who believe that humans are descended from these giant snakes. Natural curves in their local streams are attributed to an ancient python god who slithered along the surface of the earth. Ms. Coulson said, "The snake symbol runs through all the mythologies, stories, [and] cultures of southern Africa."

Inside the cave, scientists found hundreds of spear-heads, but no animal bones, cooking implements, or evidence of fires used for cooking food. The absence of such implements for daily activities suggests that the cave was used for rituals. Furthermore, within the cave is a curved inner chamber, suggestive of a snake. The archeologists theorized that this chamber could have been a place for a shaman to prepare for ceremonies. Through the archetypal symbols he etched into the rock, your ancestor left behind tangible evidence of his soul's quest for myth and meaning. Though crude in comparison to the spiritual artwork of later civilizations, this Stone Age serpent carving serves as a powerful reminder that myth connects us to the deepest parts of ourselves in ways that logic and reason alone cannot.

Such artifacts as the python carving bear witness to a powerful biological instinct that goes beyond our previous understanding of early man's needs. Something in your ancestor's DNA drove him to such specific behaviors as:

- Marking off territory for ceremonial purposes,
- Gathering in groups in these marked-off territories,
- Participating in activities that did not directly contribute to daily survival,
- Creating and constructing artwork using archetypal symbols, and

- Seeking counsel from a shaman who could receive and interpret messages from the unseen world and the world of nature.

the role of a shaman

Archeological evidence points to shamanism as the oldest spiritual practice on the planet. Tribal groups and communities drew together to receive counsel and guidance from a man or woman whose gift was spiritual rather than physical strength. A shaman had given knowledge of which plants could heal and which could kill. He was trained by his predecessor in such matters as entering into altered states of consciousness, lucid dreaming, and communicating directly with the world of spirit. Through divination, a shaman acquired information about how to prepare special offerings for the spirits, who would then guide the hunters to find abundant herds. In later years, that knowledge expanded to include protection for that season's crops. The shaman was tasked with keeping the myths of Creation, Life, and Death. Ancient and contemporary shamans believe that all of nature speaks. It is the shaman's task to interpret the messages in air, water, earth, and fire.[2]

The recent growth of curiosity about different spiritual traditions has led to a rebirth of shamanism in Western industrialized nations. But you will not find "shaman" in the U.S. Government Handbook of Occupational Titles. Despite the current interest, shamanism remains a traditional practice that is common among indigenous cultures around the world. "In shamanic communities, most messages are considered to come in forms more subtle than spoken voices…through life's synchronicities, through climate change, even through disease," writes Eve Bruce, MD. A plastic surgeon from Maryland, Dr. Bruce describes her journey of exploration in *Shaman M.D.: A Plastic Surgeon's Remarkable Journey into the World of Shapeshifting*. A person comes to understand that he or she is being called to this work through a series of tests, lucid dreams, and synchronistic events that cannot be ignored any more than your ancestor could fight the instinct to carve those symbols into the cave wall.

As scientists continue to learn more about the biological nature of spirituality, the DNA of the shaman may well turn out to contain an impressive amount of information. As some individuals possess mathematical or musical genius, there may be those whose intuitive and spiritual talents are similarly encoded in their genes.

from elemental religion to monotheism

Your ancestors' survival hinged on coexistence with the natural world. Therefore, it is not surprising that individuated soul-archetypes of deities—a god of fire or a goddess of the river, for example—would emerge as a spiritual intermediary to facilitate human interaction with the supernatural realm.

As a citizen of the 21st-century industrialized world, such spiritual beliefs may seem infantile to you. But prayers to individuated soul-archetypes are part of millions of people's spiritual practices. In Roman and Orthodox Catholicism, such soul-archetypes are the saints. In Hinduism, they are called gods.

Monotheistic religions—Judaism, Christianity, Islam, and Zoroastrianism—are based on the fundamental belief that there is one Supreme Being. (Saints, angels, and guardian spirits are believed to serve the Supreme Being by acting as intermediaries between Him and humanity.) Monotheistic religions share common beliefs that this Supreme Being has absolute power, absolute knowledge, and a conscious purpose for each individual human and for humanity as a whole. As portrayed in the Bible and the Koran, the Supreme Being, called God, expresses wrath and delight at the humans. Eastern religions—Hinduism, Taoism, Buddhism, and Confucianism—believe that the Supreme Being is an all-loving invisible force of limitless compassion, devoid of human personality traits.

continuity of life

One of the core teachings of all spiritual traditions is belief in the continuity of life. When our physical existence ends, our essence or spirit leaves the physical body and enters the Afterlife. At this point, religion becomes murky, as some preach hellfire and damnation for anyone who is not "one of them." Other traditions teach that after death, the soul leaves the physical body and journeys through states of consciousness in which it examines what it learned during its human lifetime. Many religions believe that the soul reincarnates into human existence again and again, each lifetime serving up opportunities to grow, change, and learn. How one progresses after death depends, in part, on one's consciousness during a given lifetime.

Wherever you go after this existence and whatever happens is a mystery. Your beliefs may well provide a sense of comfort, but the experience of what happens next remains unknown. Whether you believe in heaven, hell, a bardo state, or simply nothing, there is no MapQuest for the Afterlife (or absence thereof). Despite the universal themes and similarities, anecdotal accounts of near-death experiences may be relevant for those individuals who describe leaving their physical bodies and ascending through a tunnel to a beatific white light, only to return to their human selves. But does that really happen when you die? Nobody knows.

Where do such beliefs originate? Are they an aspect of our spiritual instinct? The earliest burial rituals that point to a belief in life after death began somewhere between 50,000 and 30,000 years ago. The charred remains of animal sacrifices discovered at crude altars in high mountain caves in Europe and Asia provide further evidence that Stone Age humans had some sort of belief system in the continuity of life.

When your sloop-browed ancestors were put to rest, their graves were carefully provisioned with tools, weapons, clothing, and other essential supplies for the mysterious journey ahead. The body was placed in a grave carved to represent a woman's vagina, the portal through which birth occurs. This symbolized the person's soul being reborn on "The Other Side." Cowrie shells and objects made of bone and ivory were placed alongside the corpse with a supply of red ochreous powder, the color of blood, symbolizing the life force. Your ancestor believed that these objects would be useful in the next life. Even the blood-red powder would be beneficial as a body paint to reinvigorate the new soul-body.

evolution of consciousness

As humanity evolved from being a hunter-gatherer species to one that was capable of cultivating the land for food, human consciousness shifted. With improved planting and harvesting techniques, agriculture made food supplies more predictable. As in earlier times, people banded together to form groups and then communities. Living, hunting, and farming together gave them greater protection from predators and other threats to their survival.

The agricultural revolution began around 11,000 years ago in the Middle East. (The great Egyptian civilization came into being some 7,000 years ago, around 5000 BC.) A human living prior to 9000 BC is believed to have had no awareness of his or her inner world. Everything he or she experienced was projected outside the self, toward nature. There were no thoughts or feelings as you understand them. Nor was there a sense of self or any subjective perception of life. As the species evolved, females developed the ability to hear their children's "separation call," a specific set of sounds made by mammalian offspring that triggered the protective maternal instinct.

As human consciousness changed, human beings began to develop thoughts, feelings, and an awareness of being aware. This interiorizing process became known as the development of consciousness. No longer was your ancestor limited to visceral physical survival.

Life in a community was more demanding than the survival lifestyle of hunter-gatherers. Stone Age humans did not think, nor did they have language. Governed by their primal needs for safety, their conversations, if you could call them that, consisted of grunts and guttural sounds whose messages were intended to warn others of danger. Language evolved so that people could express concerns that were more complicated than in earlier times. Planning and organizational skills evolved, along with the ability to conceive of a future measured by periods of daylight, seasons of rains, and cycles of the tides and moon.

the birth of god

As communities grew into larger, more complex cultures, certain individuals began to emerge as leaders. Members of a community would gather into groups to listen to the leader give advice, commands, and instructions. When a leader died, it is believed that his or her people were able to hear the leader's voice as an internalized command. In *Structures of Christian Existence*, John Cobb hypothesizes that such internalization of a leader's voice could have been the first conceptualization of God.

This brings us to one of the formative chicken-egg questions: Was this giant shift in self-awareness and human consciousness a result of biological evolution? Or, as recent findings in molecular biology suggest, was this shift in human consciousness an environmental trigger that led to biological changes in human DNA?

your spiritual instinct

Evidence from the New Biology shows that even before language existed to describe it, a biological instinct for spirituality had been wired into your ancestor's physiology. Had he been prevented from expressing that instinct to carve a python into the rock wall of that cave in Botswana, he may have fought to protect his right to act on that instinct, just as he would have fought to protect his food or shelter.

The word *spirituality* derives from the Middle Latin word *spīrituālitās*. It came into use during the Middle English period, sometime between 1375 and 1425.[3]

Spirituality is defined as:

- The quality or fact of being spiritual.
- Incorporeal or immaterial nature.
- Predominantly spiritual character as shown in thought, life; spiritual tendency.[4]

Since the last quarter of the 20th century, spirituality has become such a popular topic that the word has slipped into colloquial use. In the 1960s, *spiritual* referred to the gospel music of African-American religious services. *Spiritualism* referred to a religion that espouses after-death communication. The concept of individual spirituality was not yet in vogue. Now you can walk into any library or book store and find hundreds of books, CDs, and DVDs on this subject.

Spirituality is a deeply personal, often-mysterious internal process. Unlike any given religion, spirituality is not necessarily systematic, and can be expressed privately or with others. Religion is an organized community of people who share spiritual beliefs, rituals, and ethics by participating in a faith and/or organization. If you have ever met someone on *Match.com* who describes him- or herself as "spiritual but not religious," you have discovered this is an individual whose practices probably do not include worshiping at a church, synagogue, or mosque on regularly scheduled occasions. Eclectic choices of prayers and rituals from a variety of religions frequently characterize a "spiritual but not religious" person.

In *The Fourth Instinct: The Call of the Soul*, Arianna Huffington writes, "Religion is man's response to God; we come together, we bow our heads, we celebrate in song and prayer. But religion is also God's response to man. When religion is alive and vital…it helps us reconnect with God, with the spiritual reality in all things… The Fourth Instinct longs for an active relationship with the divine."

In the context of the New Biology, spirituality can be redefined as a genetically influenced belief in a supernatural entity (the Universe, Source, God, and so on), and a code of ethics based on values such as compassion, love, integrity, and selfless service. Both spirituality and religion have codes of ethical conduct that govern intentions and actions toward self and others. However, ethics address issues of respect among humans, whereas spirituality and religion address an individual's relationship to God, Spirit, or the Universe.

spirituality, religion, and the birth of altruistic ethics

With the evolution of community came values: truth, integrity, freedom, service, and sacrifice for the common good. You may not agree that these specific values are necessary for your personal happiness, but your values influence your perceptions of what creates or detracts from experiencing joy.

Ethical principles and altruism developed as people in communities recognized they needed to care for each other in order to flourish. Growing beyond the need for physical survival, humans groped for a set of principles and values that would give their lives a sense of purpose. Like the coils of the python carved into a rock in Botswana, altruistic ethics, compassionate action, service, integrity, and the pursuit of truth became intertwined with such universal principles as not killing and doing unto others as you would have others do unto you.

From this reservoir springs true happiness.

questions for the reader

- What are your impressions of your Stone Age ancestor?
- How have your impressions changed as a result of reading this chapter?
- What do you believe occurs after death?
- What places and activities do you consider "spiritual"?
- Would you consider yourself to be "spiritual," "religious," or neither?
- Why do you agree or disagree with the model of a spiritual instinct?
- How would you feel if you were prevented from expressing something important to your soul?
- What is your favorite myth? Why?

chapter 5

Dimensions of Spirituality

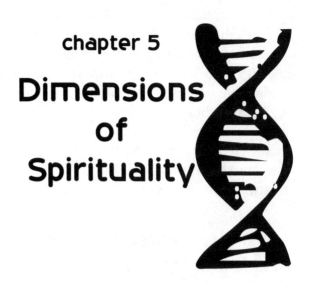

There are two excesses: to exclude reason, and to admit only reason.

—Blaise Pascal

Whether or not you believe in God, there exists within each of us an ancient spiritual instinct. Even your Stone Age ancestor needed a way to connect with an indefinable, invisible power that he recognized as being greater than himself.

You can experience a soul connection to a place, a person, or even a pet. Knowing that you are completely accepted and loved unconditionally in that place, or by that special one, results in a joy that is a natural expression of spirituality. Our human needs for love, happiness, kindness, acceptance, and connection are universal and cross-cultural. You can nurture those needs wherever you happen to be, such as in a house of worship, a beautiful garden, or even waiting in line. You can feel it on a sunny day, on a moonlit night, or sitting next to a fire listening to a storm swirling around your home. You can hear it in silence and in music, feel it when you offer comfort to someone who is hurt, and see it in your child's smile. In gratitude, humility, compassion, or sheer desperation, you have probably called out to your God during a particularly intense event. Whether you recognize it or not, your spiritual instinct is essential to your happiness. Like gravity, it exists whether you believe in it or not.

a new biology of spirituality

Scientists have begun investigating the roles your genes play in spiritual experience. Boston University neurologist Patrick McNamara has been exploring the role of dopamine, a neurotransmitter that could be involved in such spiritual feelings as a connection with God or self-transcendence—the latter being the feeling of absolute unity with all of Creation that has been described by mystics ancient and modern, from St. John of the Cross to St. Theresa of Avila to *New York Times* best-selling author Caroline Myss. According to McNamara, genes that regulate the flow of dopamine to the limbic system (emotional brain) and prefrontal lobes of the brain "support all sorts of complex functions...related to religiosity...supernatural agents and engaging in rituals."[1] Sometimes called "the heart of the mind," the limbic system is the site where emotions originate and where emotional memories are stored as molecules.[2] Mystics who report feeling joy, self-transcendence, and ecstatic union with the Divine are describing limbic states of feeling. Scientists have identified the prefrontal lobes, responsible for complex processes such as reflection, as the areas of the brain activated during meditation and prayer.[3]

In a study of 200 men in Duarte, California, conducted in 2000, genetics researchers from the City of Hope Medical Center discovered that "people with a particular variation of the DRD4 gene scored higher on the self-transcendence scale."[4]

But lest you think that specific genes are necessary for you to experience your innate spiritual happiness, let's not forget the impact of epigenetics/environment on how your genes express themselves. To reiterate this point, Ron Cole-Turner, a bioethicist at Pittsburgh Theological Seminary, said, "Genes don't create hardwired, deterministic outcomes. It's by no means determined that if you have a certain gene sequence you are spiritual and if you don't have a certain gene sequence you can't be spiritual."[5]

what gives your life meaning?

Before you dismiss this chapter as irrelevant to you, consider what gives your life value. Is it friends? Family? Your pets? Work? Your favorite form of exercise? Reading? Listening to music?

When you engage in an activity that fills you with a sense of connection to something greater than yourself, or you feel uplifted, peaceful, or happy as a result of engaging in that activity, it is a form and/or function of spirituality. When you are upset, where do you go to feel calm? Are you drawn to a place in nature where you always feel at peace? Your spiritual instinct is leading the way.

Do you attend a church, synagogue, or mosque on a regular basis? Do you choose to pray with others on a regular basis because it helps to release everyday tension and center yourself? Do you practice meditation, alone or with others?

Whatever your path and your preferences, realize that you do not need to follow the rituals or beliefs of a religion for your spirit to be joyful.

religious traditions

For the past three millennia, Judaism, Christianity, and Islam have worshiped a monotheistic God. But that was not always the case. Early mankind conceived of and worshiped a variety of gods that were eventually replaced by even more powerful deities that met their needs.

Although other religions, such as Hinduism, hold to a pantheistic framework that embraces gods, saints, and spirits, their purpose is to serve a Supreme Being by functioning as intermediaries.

the times, they are a-changin'

Prior to the Age of Reason, which began in 17th-century France, populations were terrorized and tortured into believing in God. During the Crusades and the Inquisition, questioning the authority of the Church was punishable by torture and death. The Age of Reason established that man had a divine right to think and gain awareness of his individual identity. Such advances in thinking enabled man to develop scientific methods of inquiry about the natural world.

Mathematicians and philosophers in the Age of Reason separated the physical universe from the realms of spirituality. They believed that these worlds did not interact. This concept was accepted by the Church, which stopped persecuting intellectuals for having different ideas about reality. And for more than 400 years, scientists refused to study anything that was not material, believing that it was a topic suited for theologians and philosophers.

Darwin and "The Big Debate"

In the 19th century, a British naturalist named Charles Darwin further revolutionized Western thinking with a theory of evolution that challenged the biblical version of how God created the world. Charles Darwin's Theory of Evolution was based on scientific inquiry and biological evidence. As the Industrial Revolution advanced, scientific thinking led to new inventions that changed the course of civilization. Along the way, a chasm opened up between the priests of science and the priests of religion. In European countries, the United States, and Canada, logical, linear thinking sought to replace faith-based ideas. Spirituality, intuition, and emotional intelligence were considered less effective than rational thought. Yet we must bear in mind that it was none other than Blaise Pascal, the 17th-century French mathematician known as "The Father of Reason," who cautioned: "There are two excesses: to exclude reason, and to admit only reason."

By the middle of the 20th century, Darwin's theory was universally accepted as the science of biology and taught in a majority of American schools. However, the end of the 20th century was marked by intense ideological debates between religious factions wedded to the literal biblical version of creation and those who believed in evolution.

While this debate still continues today, it has been reduced to a minor controversy simmering around small centers of special interests. The major religions of today accept the science of evolution. Even the conservative Catholic religion has taken the position that it doesn't disagree with the Theory of Evolution. Even so, they don't talk about it, and consequently much of their non-scientific laity prefer to adhere to tradition rather than evolution. Because the science of biology is based on the Theory of Evolution, it is just a matter of time until the Theory of Evolution will be commonly accepted as the way it is.

assumptions of Western science

Three primary assumptions of Western science have become imprinted into the collective consciousness:

1. **Reductionism:** Only that which can be reduced to matter is real. Atoms and molecules are the substance of "the real world."

2. **Objectivism:** Only that which can be studied objectively is real. Subjective experience is not real and does not count.

3. **Positivism:** Only that which can be measured is real.

According to these assumptions, spiritual states of consciousness cannot exist because they are subjective and cannot be measured.

Yet if you tried to measure water in a sieve, would you call the water an "artifact" and deny its existence? Or would you design a new way to measure it?

Is your consciousness real, even though it is immaterial?

challenging the assumptions of science

In the late 20th century, the New Science movement began to challenge those outdated assumptions. Today, new findings in quantum mechanics reveal that one can *never* specify the state of a subatomic particle, such as its simultaneous location and velocity, with complete certainty. In other words, all that is assumed to be material cannot be measured with certainty. Thus, new discoveries in biology and physics called into question the standard reductionist model of traditional Western science.

Since the late 20th century, science and reason have become a religion to millions of educated individuals who have forgotten that "to admit only reason" is as egregious as excluding it. Disregarding Pascal's prescient words of wisdom, today's rationalists argued that only what is logical and sensory-based is true. But what we call "science" is only some 300 years old. It is a relatively new language that describes humanity's study of the universe. Many of the world's spiritual and indigenous traditions have been around for thousands of years, and had their impact on human thought and behavior influenced by the epigenetics of environment.

Gregg Braden, author of *The Spontaneous Healing of Belief*, asks: "What is that space, that emptiness, that flows between subatomic particles, connecting everything in the universe?" Braden identifies the first false assumption of science as labeling the space between physical things as "empty." He says, "These scientists tell us that about 96 percent of the universe is empty. Only about 4 percent contains what we would call stuff—physical matter."[6] But we now know that this "emptiness that flows between subatomic particles" is a universal field of what physicists refer to as *non-conventional energy* that underlies everything we see in our physical world. Modern science is really struggling to come to terms with this invisible field. For years, scientists have been arguing about the nature of whether or not it exists. Whereas they have only recently come to accept that the field is real, indigenous and spiritual traditions thousands of years old *began* with this understanding.

"We're all connected through an energy field or matrix," says Braden, who refers to it as "the Creator, the Directing Intelligence."[7] In *The G.O.D. Experiments: How Science is Discovering God in Everything, Including Us*, Gary Schwartz defines G.O.D. as an intelligence that is constantly "Guiding, Organizing, and Directing." Both Schwartz and Braden concur that humans and G.O.D. can converse.[8] According to Braden, "We can communicate with this field, this intelligence, if you will, through the nonverbal, the kinesthetic language of the heart. That is the essence of prayer."[9]

conclusion

Whether you believe in God, G.O.D., or only what appears logical and measurable is irrelevant. A sense of connection and empathy for all living creatures is essential for the evolution and survival of the human species. When your ideas, emotions, and values lead to concerned and compassionate behaviors, your level of happiness rises. So do the happiness levels of others around you. You may be surprised when, after a while, your baseline emotional state is less concerned with the wounds of the self because you experience yourself as connected to other people, the environment, and the Universe itself.

questions for the reader

- Do you believe in God? G.O.D? Neither? Why or why not?
- Have you experienced a sense of connection to "The Universe?"
- When? What were you doing at the time?
- List three to five beliefs that sustain you in times of crisis.
- Describe yourself as a spiritual being.
- What moves you? What gives your life a sense of wholeness?

chapter 6
"Common" Happiness

Since World War Two, we have become stuffologists. We believe the gathering of stuff is a show of intelligence. Our nation's soul is on a respirator.

—Caroline Myss, PhD

Have you ever bought something—anything—thinking it would bring you happiness?

What happened?

After the initial rush of pleasure, did you find that owning or using that car/lipstick /iPhone/ShamWow elevated your mood to the point that you could say, "I am now filled with joy because I have my [object]?"

I'm not suggesting that there is anything wrong with buying something that enhances your comfort or quality of life. However, we do believe that the relentless pursuit of happiness through the acquisition of objects is like a sugar high: a quick rush, then a drop in mood, followed by a new craving to buy something else.

97

There is a multi-billion dollar industry dedicated to hypnotizing you into believing that unless you buy a particular thing you won't be successful, popular, or sexy.

are you a "stuffologist"?

An informal survey of more than 500 executives found that more than half of them had made a major purchase because they had seen a seductive commercial on television, only to find themselves disappointed or frustrated by the result.[1] That SUV on top of the mountain lured many with a dream of spending more time outdoors in rugged natural surroundings. But when they got it home, they found that it was impractical for their daily use. Some said it did not fit into the garage at home or at work. Others were disappointed by the low gas mileage. Everyone in this informal survey sheepishly admitted to allowing themselves to be swept away by the idea that buying one of these vehicles would significantly improve their quality of life.

> *How hard it is for the wealthy to enter the kingdom of God. It is easier for a camel to pass through the eye of a needle than for a rich man to enter the kingdom of God.*
> —Mark 10:23–31

For centuries, people have been asking whether money can buy happiness. We don't think this can be answered with a simple yes or a no. Taking a friend to lunch, splurging on a birthday gift, or going on vacation with your BFF can briefly lift your spirits and generate the kind of emotional memories that make you smile. However, whether you live in a penthouse, a raised ranch, or a rustic bunk in the middle of nowhere, you can experience comfort, pleasure, peace of mind, and joy. For instance, when traveling through a desert in northern Venezuela, Laurie learned that even people who live in concrete huts without indoor plumbing can sometimes be happier than those of us with Jacuzzis and air conditioning. The Venezuelan desert people love to party. At night, they turn up their boom boxes, grab a few beers, and dance under the stars. "What else do we need?" one of them asked rhetorically. (You might want to ask yourself, as Laurie did, "When was the last time I danced under the stars?")

Your values, relationships, and beliefs can make or break your baseline emotional state. A University of Rochester team of psychologists studied a group of young adults during their first year after graduating college. Researchers found that those recent grads whose needs for autonomy, competency, and connection to others were met, were simply happy. Those who focused on achieving extrinsic goals—earning a good income while doing a job they hated—were, frankly, miserable.[2]

...but rich people have more sex!

If you earn more than $9 million a year and have $89 million or more in net assets, you are more likely to have expensive sexual tastes. In a 2007 study of "Money as Aphrodisiac," Hannah Shaw Grove and Russ Alan Prince interviewed 661 men and women of high networth about their sexual preferences. More than half of those surveyed reported that they believed that their private wealth gave them better opportunities to pursue an optimal sex life. It is worth noting that 57 percent of respondents had been divorced at least once. Nearly all the participants said that money freed them up to travel to foreign destinations and engage in social activities that were oriented toward adventurous or exotic sex.[3]

There is no question that you will be happier with a fulfilling sex life. Grove and Prince notwithstanding, your sexual needs change with age, hormonal shifts, and maturity. Whether you have a high networth or one that is considerably more modest, in time, your sexual fulfillment is more likely to evolve with a long-term or lifetime partner. Whether you are married or have a lover, are heterosexual or homosexual, sexuality develops into a way of expressing how you feel about each other. A long-term sexual relationship that is caring and respectful is more likely to become a "happiness asset" that gains value as you both grow and change.

what is a "happiness asset"?

Make a list of the people, activities, places, music, movies, books, food, pets, and situations that bring you happiness.

On a scale of 0 to 10, in which 10 is the peak or maximum level of joy you experience, rank each "item" on your list.

Review your list with a friend or family member. Discuss how each of these brings you happiness and whether your happiness is temporary, sustainable/consistent, or increasing in time.

Anyone or anything that brings you happiness that increases in time is one of your "happiness assets."

As with any other asset—your home, jewelry, artwork, and investments—your happiness assets have special value to you and should be treated accordingly. Never take your happiness assets for granted! Care for them as you would care for your home or financial investments, for your happiness assets will continue to bring you increased happiness if you value and take good care of them. At the end of the day, you will probably find that nurturing your happiness assets brought you deep and lasting joy because they responded and reciprocated in kind.

happiness by gender

The annual U.S. General Social Survey asks both men and women how happy they are on a scale of 1 to 3. Three means "very happy"; one, "not so happy." More than 1.3 million people have participated in the survey since it began in 1972. The most recent findings, from the 2007 survey, show that women report being happier than men when they are younger. With time, women's reported happiness declines, whereas men's increases, starting around the age of 47. This is due, in part, to women's losing their extrinsic value in American society as they become older and are considered less attractive in the culture as a whole than when they were younger. In a society that prizes youth, beauty, and health, women struggle with self-esteem issues at a time in their lives when they would be respected for their experience and wisdom in a different culture. Ironically, the 2007 survey shows that despite—or perhaps because of—the greater opportunities available to women compared to the early 1970s, women describe themselves as less happy than they did in 1972.[4]

According to the World Health Organization, the increasing responsibilities that accompany greater freedom and opportunities for women carry a psychological price tag, and greater stress. Stress is blamed as a primary cause of depression among women globally. After heart disease, depression ranks as the second-most-serious illness for women worldwide. It ranks as #10 for men. These statistics may represent a significant difference in gender styles with regard to reporting psychological distress: As a cross-cultural phenomenon, women are more likely to admit they are depressed and seek professional help than are men.[5]

world happiness map

In the Netherlands, a "continuous register of scientific research on subjective appreciation of life," includes data collected from 4,252 national surveys and 1,220 instruments measuring happiness in 155 different countries.[6] The data about how individuals in each country listed appreciate life has been collected during national surveys similar to the U.S. General Social Survey.

This project started in 1993, when Dutch sociologist Ruut Veenhoven published "Happiness in Nations: Subjective Appreciation of Life in 56 Nations 1946–1992."

The indicators surveyed include the following themes:

- Consumption
- Cultural climate
- Crime

- Demography
- Education
- Freedom
- Geography
- Government
- Happiness
- Health
- Inequality
- Institutional quality
- Law and order
- Lifestyle
- Modernity
- Personality
- Politics
- Risks
- Social cohesion
- Values
- War
- Wealth[7]

Asking "how much people enjoy their life-as-a-whole on a scale of 0 to 10," Veenhoven and his team compiled an overview of "average happiness" in 144 nations compiled from data obtained from 2000 to 2008. The top-scoring (happiest) nations may surprise you:

- Iceland
- Denmark
- Colombia
- Switzerland
- Mexico

Ranked in the middle are:

- Philippines
- China
- Iran
- India
- South Korea

The lowest scores on the planet—meaning the least-happy countries—were reported as:

- Chad
- Togo
- Angola
- Zimbabwe
- Tanzania

These rankings were based on the average scores, with the top range being greater than 7.9, the middle range plus or minus 6.0, and the bottom range greater than 4.3.

The United States and Great Britain came in with 7.0 and 7.1, respectively, indicating that Americans and Brits are happier than Filipinos but not as happy as Mexicans.

Produced in 2007, the world map of happiness is based on studies conducted by UNESCO, the CIA, the New Economics Foundation, the World Health Organization, the World Database of Happiness (Veenhoven), the Latin barometer and Afro barometer, and the United Nations High Commissioner for Refugees. Its author, Adrian G. White, is a psychology professor at the University of Leicester in the United Kingdom.[8]

The 20 happiest countries out of a total of 178 are:

- Denmark
- Switzerland

- Austria
- Iceland
- The Bahamas
- Finland
- Sweden
- Bhutan
- Brunei
- Canada
- Ireland
- Luxembourg
- Costa Rica
- Malta
- The Netherlands
- Antigua and Barbuda
- Malaysia
- New Zealand
- Norway
- The Seychelles

The United States ranks #23; China, #82; Japan, #90; and India, #125. The United Kingdom ranks #41 among 178 nations.

The discrepancies between White's map and Veenhoven's database could be due to differences in the source materials used to generate the data. In describing his study of happiness as "intangible but important," White observed, "A nation's level of happiness was most closely associated with health levels (correlation of .62), followed by wealth (.52), and then provision of education (.51).[9]

The 18th-century British philosopher Jeremy Bentham contended that the purpose of politics was to bring the greatest happiness to the greatest number of people. A recent British survey shows that 81 percent of the population agrees with Bentham that the government should concern itself with creating happiness rather than wealth. One politician commented, "It's time we admitted that there's more to life than money. It's time we focused not just on GDP, but on GWB—general well-being."[10]

conclusion

In guaranteeing U.S. citizens the right to "life, liberty, and the pursuit of happiness," the Founding Fathers were not thinking of the powerful subliminal influence that commercial advertising would have on future generations of Americans.

Despite our comfortable standard of living, access to information, and civil freedoms, Americans are far from the happiest people on earth due to addictive pursuits that provide a momentary rush of pleasure, followed by disappointment. This, in turn, leads to a new cycle of craving the next object du jour.

Common happiness is based largely on extrinsic circumstances—the belief that something external to us can provide a consistent source of happiness. But that is not the case, for the nature of our circumstances is change.

In looking at the research, don't you wonder what isn't working here?

Why are we somewhat happier than Filipinos and less happy than Mexicans?

Is it possible that people in other parts of the world know something that we, in our technologically sophisticated world, do not? Are the happier cultures more stable, with less change of circumstances? Are they more willing to accept "what is" rather than constantly change?

Are there skill sets for happiness that you can learn? If so, what are they, and how can they make your life truly happy, thereby activating some of the genes that will help your children and grandchildren to be happier too?

Would you like to find out?

questions for the reader

- Are you a "stuffologist?" Why?
- What specific possessions make you happy?
- What is it about them that make you happy?
- What external circumstances make you happy?
- When these external circumstances change, how will you stay happy?
- Do you ever think that you spend too much time thinking about buying new things?
- Have you ever lost, misplaced, or gotten rid of something that brought you happiness?
- How did you feel after it was gone? Were you relieved? What did you do after it was gone?

chapter 7

Natural Happiness

*As we cultivate happiness in ourselves, we
also nourish happiness in those we love.*

—Thich Nat Hanh

*When one door of happiness closes, another opens;
but often we look so long at the closed door that we
do not see the one which has been opened for us.*

—Helen Keller

You might be wondering to yourself why we started
this chapter with two quotes on happiness when each of
the previous chapters began with just one quote. But if
you read them again, you may not be able to stop smiling, because, when it comes to your happiness, more is
better. You might even think, "Hmm...so easy, even a
caveman can do it."

That's right! There is nothing complicated about getting
to happiness. Yet thousands of research papers, dissertations,

studies, and pop psychology books have been written to explain in structured, logical terms what is simply a natural, spontaneous...

"A natural, spontaneous *what*?" you may be wondering.

As you begin to wonder, any number of possibilities may come to mind.

Is it an emotion? A mood? A state of mind? A state of being? A spiritual experience? A delicious sensation? An activity or event? Hanging out with your BFF? Enjoying a cup of coffee as you sit in the car waiting for your children to get out of school? An afternoon getting ready for a hot date? Your team winning the World Series? A solitary afternoon curled up with a great book? Fishing on a summer lake? Listening to your iPod as you work out? Finding time to visit an old friend? Cooking something from your favorite recipe and sharing it with family and friends?

Does happiness have a color? A sound? A texture? Does it have a size? A shape? A number? Does the size, shape, or number change or move? Or is it always the same?

Yes, these are odd questions. But they are designed to bypass your preexisting concepts of happiness by triggering your playful imagination, and hopefully a cascade of happy thoughts, feelings, and sensations.

Happiness happens, spontaneously. When it does, smile.

But what is "it"? A noun that you can't put in the trunk of your car? Hold in your hand? Pour into a cup?

the results are in!

According to Wikipedia, "happiness is a state of mind or feeling characterized by contentment, love, satisfaction, pleasure, or joy."[1] It is the word most frequently used to describe a life that is lived with unconditional love and compassion.

Research conducted by scientists at the Mayo Clinic in Rochester, Minnesota, have concluded that the means exist to quantify and measure happiness using surveys and instruments that assess well-being and its components.[2] As a result of their research, they contend that you cannot buy happiness, but as discussed in Chapter 6, you *can* buy pleasurable goods, trips, activities, and experiences. The Mayo Clinic findings show that your sense of well-being is not likely to change much with a higher standard of living.

genetics and epigenetics play an important role

An overview of studies on happiness shows that your genes are responsible for about 50 percent of your personality traits. The other half is due to your interactions with family, culture, and environment. In a word, epigenetics![3]

Do you remember the study of suicide victims who had reported being victims of child abuse? Their autopsies showed that a gene that regulated the flow of stress hormones was "methylated" so that it could not function properly. Like sticking chewing gum on a light switch, methylation prevents a gene from being activated, also known as "expressed."

Following this idea, researchers at the University of Texas Southwestern Medical Center studied the effects of aggressive neighbors on adult mice. The mice reacted similarly to the way human beings do: They became depressed due to a gene that was altered biochemically from the environmental stressors. Scientists gave the mice antidepressants, which successfully reversed the depression that was triggered by the first phase of the study.[4]

So how much of your happiness is determined by heredity (traditional genetics) and how much is due to environment (epigenetics)?

Because you are standing at the beginning of a new branch of scientific research, there is no "correct" answer or fixed percentage, in part, because "there is more to heredity than genes."[5]

As I've said, this is not what you were taught in high school or college, but cutting-edge discoveries in molecular biology contradict the traditional version of Darwin's Theory of Evolution, which believes that evolution occurs through natural selection, with instructions for inherited traits passed on exclusively through genes. As you are probably starting to realize, *this is not true all the time*.

Not only does this New Science challenge those assumptions you were taught in basic biology class, but it also leaves room for you to experiment with some of these new principles in order to improve your well-being and health, thereby creating new information and signals that your cells can transmit to the next generation of cells. As you program your thoughts, emotions, and behaviors to develop daily habits of happiness, you are creating new epigenetic patterns that alter how your genes behave.

To go back to our chicken-egg question of how much of your personality is genetic and how much is epigenetic, the research would indicate that the ratio is flexible; as far as we know, there is no fixed percentage. Understanding the basics concepts of epigenetics can give you powerful options for reaching new levels of happiness that can, in turn, carry over to new generations of cells. As you become happier, your cellular structure changes. Some of these positive changes imprint your children's genes, as well.

your expectations of happiness

It goes without saying—but I'm going to mention it anyway—that your happiness, or lack thereof, depends largely on your expectations. If you believe that only one person, job, home, or car can bring you happiness, you are setting yourself up for disappointment. Even if you fall in love with your childhood sweetheart, marry, and raise a healthy family, eventually one of you will depart, leaving the partner to grieve. Nothing in this world lasts forever. Loved ones, jobs, homes, cars, finances, and health can come and go. If you believe that living a happy life means that you will never go through loss, sorrow, disappointment, failure, or grieving, you are setting yourself up to be let down. Your beliefs about happiness need to be realistic. If you are addicted to a fantasy of perfection, any happiness you experience will be short-lived. To expect anyone's life to be flawless is utterly unrealistic.

This brings us to the next expectation: If you think happiness means that you never feel sadness, rage, or remorse, your beliefs about happiness are, shall we say, "reality-challenged." A sudden upheaval in your personal or professional life, extreme financial difficulties, continual family crises, divorce, surviving an accident, witnessing a crime, or catastrophe are likely to disturb even the sunniest disposition. It is not a failure to find that you don't feel like yourself, or that you can't shake a deep sense of sorrow after such an event. Even a seasoned first responder or combat veteran can become upset. Feeling disturbed by violence or sudden loss does not mean that you are *not* a happy person or that you will *never* be happy again. That type of statement is called a "cognitive distortion" because it leads you to an all-or-nothing, either/or, black-or-white outcome. Happy people are capable of being shaken up emotionally by tragedy. They are capable of mourning their losses and feeling the 10,000 sorrows and 10,000 joys life has to offer.

And so are you! Feeling sad or angry does not mean that you are a "victim of negative thinking" or that you are being punished...as long as you do not get "stuck" ruminating about hurtful events in the past. As a vibrant human being, you need to be open to a spectrum of emotions. That way, you can enjoy what has been nicknamed, tongue in cheek, "the full catastrophe of Life with a capital L."

diagnose and adios!

As a culture, we Americans tend to be phobic of unpleasant life events. Our superficial cultural programming conditions us to believe that we are entitled to feel joy at all times. If we are not experiencing pleasure, millions of us feel as if we have somehow failed or that God is punishing us for negative thinking. The late Margaret Mead, a world-renowned anthropologist, observed that of all the societies in the world she had studied, only Americans believed that if they became ill, it was a punishment from God.

We also have a tendency to use the words *sadness* and *depression* interchangeably. But you can feel sadness without being depressed. Feeling down, feeling blue, crying, or experiencing sorrow does not automatically mean that you are "depressed." But if your mood is depressed for days at a time and you have lost interest or pleasure in what you used to enjoy, and those symptoms (along with several others) persist for a two-week period, you probably meet the diagnostic criteria for depression and need to seek professional help.[6]

Incomprehensible though it may seem, you are also capable of feeling contradictory emotions simultaneously. Have you ever laughed and cried at the same time? Laughed so hard that you cried? Felt sad when others were celebrating? Discovered humor in a desperate situation?

My point is that sadness in and of itself should not send you to your doctor for a prescription for antidepressants to make it go away. Forget the "diagnose and adios" approach. Sadness is not a disease.

Nor is happiness the opposite of sadness, although many people have said that without sorrow, we would have no way to measure how happy we are. As we noted earlier in this chapter, it is possible to feel happy and sad simultaneously. Happiness is unique to every one of us, yet the language of happiness is spoken and understood everywhere.

Although you cannot buy it, grow it, achieve it, receive it as a reward, chase it, hold on to it, or put it in the trunk of your car, you know happiness when you see, hear, and feel it. Ironically, the more frantic you are in searching for it, the less likely you are to happen upon it. And when you do, you might not be able to recognize it within yourself. Chasing after happiness while believing that it is somehow removed from you, as in "out there," or "in the future," can make you become addicted to craving what you believe you cannot have.

But you *can* be naturally happy. Right here. Right now.

national accounts of well-being

The Centre for Well-Being in London is part of the New Economic Forum (*www.nef.org*), a self-described "independent think-and-do tank" founded in 1986 to study such issues as international debt. NEF became instrumental in having international debt put into the agendas of the G7 and G8 summits. According to the think-and-do tank's mission statement: "We believe in economics as if people and the planet mattered. We aim to improve quality of life by promoting innovative solutions that challenge mainstream thinking on economic, environment and social issues."[7]

In partnership with University of Cambridge/Well-Being Institute (UK) and the European Social Survey, the NEF wants governments to consider their citizens' well-being to be as important an asset to that nation's wealth as its Gross National Product. (This is a radical idea, is it not?)

At *www.nationalaccountsofwellbeing.org*, you can fill out a brief questionnaire that gives you a spider web–shaped graph that measures eight components of happiness and well-being on a scale of zero to 10:

- Positive feelings.
- Absence of negative feelings.
- Satisfying life.
- Vitality.
- Resilience and self-esteem.
- Positive functioning.
- Supportive relationships.
- Trust and belonging.

Your results are superimposed over the European average, which is evenly distributed at a smooth number 5 in all sectors. It is fascinating to compare your individual profile with profiles from different countries, which show correlation to the World Happiness Map discussed in Chapter 6.

There are several benefits to completing the survey:

- The chart instantly shows you which areas of your life are more—and less—satisfying.
- You can see which areas of your life could use a boost.
- You can compare your profile to others.
- You can recommend this to your family and friends so that you can see how your sense of well-being compares to theirs. If such categories as "supportive relationships" and "trust and belonging" look sickly, you can begin to address what's missing by sharing your results and

talking about this with your close or not-so-close circle of friends and family.

- You can evaluate your personal criteria for happiness in comparison to the criteria in the survey. (This opens up another area for conversation with friends and relatives.)

essentials of natural happiness

A recent meta-analysis of 850 studies on happiness confirmed that "higher levels of religious involvement are positively associated with indicators of psychological well-being (life satisfaction, happiness, positive affect, and higher morale) and with less depression, suicidal thoughts and behavior, drug/alcohol use/abuse."[8]

One of the world's leading experts on happiness is a religious leader: His Holiness the Dalai Lama. His insights speak to the human heart and spirit, regardless of nationality, gender, or religious belief: "As human beings we all want to be happy and free from misery...we have learned that the key to happiness is inner peace. The greatest obstacles to inner peace are disturbing emotions such as anger and attachment, fear and suspicion, while love, compassion, and a sense of universal responsibility are the sources of peace and happiness."[9] We are not advocating for or against religion, as that is a personal decision. As Mahatma Gandhi often said, "My religion is a private matter between God and myself."

find your flow

Like falling in love, once you have found your flow your life is changed forever. Flow is a state of effortless knowing in which your mind, body, and spirit are aligned to deliver excellence. Athletes call it "the Zone." For some, "the Zone" feels like a waking lucid dream, in which you simultaneously participate and observe.

Professional and Olympic athletes aim for "the Zone" in order to execute their personal best. In ancient Japan, master archers attained states of clarity and focus so as to achieve flow at will. Blindfolded, the archers' arrows struck the center of any target.

Mihaly Czikszentmihalyi, PhD, a University of Chicago professor of psychology, is considered the father of flow. He wrote, "It is the full involvement of flow, rather than happiness, that makes for excellence in life."[10] We would argue that "full involvement of flow" is a state of natural happiness in and of itself.

You may not notice, at first, when you find your flow, because it occurs naturally when you are doing something you love: walking, running, skiing, gardening, crocheting, or even cooking. Suddenly, all the elements come together without your having to think about them. You execute your sequence of steps so that they merge into one fluid process. Emerging from flow, your mind and body may be flooded with "e-words": excitement, exhilaration, enthusiasm, and energy.

a sense of meaning

Viktor Frankl, an Austrian psychiatrist who was imprisoned in the concentration camp at Auschwitz, took note of how those around him reacted to their brutal surroundings. Many fell into severe depression, often committing suicide by throwing themselves against the electrified barbed wire fence surrounding the camp. Others held on to hope with courage and resilience, believing that they would endure and survive.

One bitterly cold day, Frankl saw, in his mind's eye, a vision of his future self. He was standing in a warm, well-lit auditorium lecturing to a full house about his experiences in Auschwitz and what empowered him and others to survive psychologically as well as physically. In an intuitive flash of knowing, Frankl knew that he would survive the camps in order to compel the world to understand the nature of suffering. This insight led to his writing *Man's Search for Meaning*, which became an international bestseller. In this book, he describes how living with a sense of meaning can bring happiness, regardless of external circumstances.

If you think of an average news day for a TV show such as *The Insider* or *Entertainment Tonight*, how many stories featuring celebrities, wannabes, and garden-variety drama queens reveal their lives to be filled with conflict and turmoil? Alternatively, those individuals who use their fame to help others in distress stand out because they are *not* the norm. They also stand out because they exemplify what it means to live with a sense of purpose and compassion.

It goes without saying (but I'm going to say it anyway): You do not have to become a celebrity to lead a meaningful life. Nurses and doctors, physical therapists, teachers, home care helpers, and troops on the front lines serve others every day without recognition. Happiness comes from finding and acting upon your sense of purpose.

Alex Pattakos, PhD, believes that "the search for meaning is a mega-trend of the 21st century."[11] A protégé of Viktor Frankl's, Pattakos helps people around the world to find their personal vision for a meaningful life. His book, *Prisoners of Our Thoughts*, is an international best-seller.

You do not need to save the world in order to live with meaning. One of the easiest ways is to be there for the others in your life. Ask. Listen. Drive someone to the airport. Offer a wake-up call for your late-sleeping colleague. Sweep your next-door neighbor's sidewalk when you are cleaning your own. Keep an extra box of tissues, hand wipes, or umbrella in your desk drawer to be shared when needed. It's that simple.

kindness and forgiveness

Holding on to anger for years at a time can block your positive emotions. No matter how hurtful a particular event in your life has been, numerous studies show that keeping a grudge is like drinking poison and expecting the *other* person to die: You are the only one who suffers. When you are unwilling or unable to forgive, you put yourself at greater risk for heart disease and lowered immune functions. Your anger can make it more difficult for your body to fight off physical illness. You also put yourself at risk psychologically. People who cannot forgive suffer from depression, anxiety, and numbness to violence. They are more likely to develop addictions to drugs and alcohol than those who forgive themselves and others.[12]

A case in point: Immaculee Ilibagiza was a college student in Rwanda when her neighbors joined in the Hutu massacre of the Tutsi people. A neighbor hid her in a tiny bathroom, where she huddled with eight other women during a killing spree that left more than 3 million people dead, including most of Ilibagiza's family. Outside the bathroom window, people seeking to kill her called her name. She did not think she would survive and spent most of her waking hours in prayer. Several years after the genocide, Ilibagiza located the man who had killed her mother and brother. He was in a prison in her home town. When the jailkeeper brought her mother's killer to see her, Ilibagiza said, "I forgive you." The jailkeeper was furious and accused her of disrespecting her family's memory. She told him, "Forgiveness is all I have to give." She went on to write a *New York Times* best-seller about her personal journey to forgiveness: *Left to Tell: Discovering God Amidst the Rwandan Holocaust.*

One of the most-researched topics in psychology, spirituality, and health, forgiveness has been found to improve cardiac and immune function, boost overall health, and reduce levels of anxiety and depression. People who forgive cope more effectively with stressors in their lives. They report higher levels of optimism and hopefulness than those who are unable to let go.[13]

A pioneering study by the Stanford Forgiveness Project found that adults who were taught such skills as meditation, guided imagery, restructuring thoughts so as to take less offense, giving up blaming their respective offenders, and beginning to understand the situation through the eyes and mind of the offender, reported lower levels of stress and greater well-being after the six-week training program by Dr. Fred Luskin.[14]

If you are unable to forgive someone who has hurt you, and you would like to take an anonymous first step, go to *www.forgivesomeone.com*, a Website that gives you a confidential format to express your feelings.

Should you wish to learn how to become more forgiving as an everyday practice, We recommend the following guidance from Pope John Paul II: "See everything. Overlook a great deal. Improve a little."

fun

Do you suffer from Fun Deficiency Syndrome, also known as FDS?[15]

"Fun is FUN-damental," says Steve Bhaerman (aka Swami Beyondananda), a self-described cross between Ram Dass and Hagen-Dasz. (Ram Dass is a former Harvard professor who dropped LSD with the late Timothy Leary and who later sought enlightenment without drugs.) The Swami and Ram Dass question some of society's "sacred cows," meaning absurd assumptions. Both believe that spirituality is no use to anyone if it isn't fun!

So find something funny every day. Stuck in traffic? What a great opportunity to listen to your favorite comedians! Stick out your tongue and make faces at yourself. When you notice the expressions of people in nearby vehicles, you probably won't be able to stop laughing.

Lost your job? Rent *A Thousand Clowns*. This 1960s movie starring Jason Robards, Jr., is the story of Murray Burns, an unemployed comedy writer who is having too much fun to look for work. As you watch him and his nephew Nick playing hooky from life, you can't help laughing. Even when you are down and out, you can still tickle your funny bone. Or, as Murray puts it: "You have got to own your days and name them, each one of them, or else the years go right by and none of them belong to you."[16]

conclusion

Happiness cannot be bought, given, stolen, pursued, or doled out as a reward. Paradoxically, the harder you try to contain it, the more elusive it becomes. However, just as you prepare the soil in your garden so that you can plant seeds that will flower for you, you can grow habits that promote sustainable happiness.

questions for the reader

- What if you could remember three times when you were happy?
- Do you recall those happy times as images, sounds (words or music), or feelings in your body?

- Recall the most recent time you have shown someone kindness. What is the most recent time you have received kindness? Which makes you happier?
- How would you begin to forgive yourself for a mistake?
- What would you say to someone whom you have not yet forgiven?

Part III:

Making

Happiness Happen

chapter 8

Getting on the Path to Happiness

Give and it will be given to you. Good measure, pressed down, shaken together, running over, will be put into your lap.
—Jesus Christ, Luke 6:38
(New Revised Standard Version)

The purpose of this chapter is provide the background and fundamentals for the Natural Happiness 28-Day Program (detailed in the next chapter), and to set the stage for the deeper happiness habits practices that follow in Chapter 10.

The path to happiness *is* clearly marked, but you need to pay attention in order to find it. In yesteryear, the signs were written with considerably less knowledge, compensated by more mystery. Today, the research and new concepts of New Biology shine a revealing light on your nature, evolution, and the universe.

The Human Genome Project now gives scientists the code to interpret how our species works, thinks, and behaves. The blueprint of our DNA can help us understand the nature of human spirituality as a source of happiness. This new information reveals an evidential link between biology and our spiritual nature.

The subject of nature versus nurture has long been a debated, but we now have an understanding of our genetic nature; with the new science of epigenetics, the mechanics of nurture become clearer. As the embryonic science of epigenetics advances, it will provide a means to speed up the evolutionary process and consciousness development. Multi-disciplinary approaches that integrate quantum mechanics, genetic engineering, astrophysics, evolutionary biology, consciousness, and epigenetics have the potential to heal our planet, which is haunted by violence and threats of ecological destruction.

Statistical polls and studies notwithstanding, instant gratification, competition, materialism, and greed do not produce happiness. To the contrary, the ancient instincts for power and aggression result in suffering, violence, and destruction in our world. But from an evolutionary perspective, we humans are simply doing what we have been programmed to do from the beginning.

Given the situation, is it realistic to think we can change our behavior from greed to charity, from hate to love, from self-serving to altruism, from materialism to spirituality? It is all very interesting to talk about being more altruistic, but how can we walk the talk? Countless "self-help" books and articles have promised a better quality of life, but obviously failed in their promise, because they keep pouring into the market. Is there any reason that this book will be any different?

We believe it is, for the following reasons:

- We humans are genetically spiritual, and are rewarded with joy when we choose ideas, emotions, and behaviors that express our "happiness genes."

- Our happiness genes work 24/7 to provide the emotional rewards that encourage the continual behavioral choices that reinforce them.

- Instincts and emotions that restrict certain regulatory genes can be reprogrammed.

- The exercises in this book are fun. This makes them emotionally rewarding, which increases the motivation to practice them consistently.

- From consistent repetition, we build habits of natural happiness that can have a positive impact on certain genes. (There is research to support this!)

- Whether or not you are moved by scientific evidence, we believe that if you practice these exercises, the results you get will provide you with sustainable happiness that will give you all the evidence you need that this program really works.

It's all about change

You may think that in order to be happy all you need to do is pursue pleasure and avoid pain. But the central message from all the great spiritual and religious teachers is that we can do better. For example, the basic message of Jesus Christ is to love God and your neighbors; some see God as the space between objects in the universe, as well as in the space between atoms in our body[1]; others believe that God is the essence of consciousness; the core message of the Buddha is to detach from your desires to end suffering. The tenets of other great spiritual and religious leaders and teachers is similar: Choose compassion for others over materialism, and happiness will be yours.

Though it is easy to teach, behavioral change is hard, as vividly demonstrated by the overwhelming failure of virtually all the self-help and diet books. The reality is that, in spite of our soaring scientific and technological advances in America, our state of happiness is declining. For example, despite constant warnings from health organizations and widespread media, we are well on our way to becoming the first developed nation to have a population of which the majority is overweight or obese. Not only does this trend cause one of our largest threats to good health, but it also reduces quality of life, which means less happiness.[2]

It is human nature to pursue happiness, but we often mistake pleasure for happiness. As is often said, the only constant in life is change, but in this case the change is in life situations. So as the circumstances of our pleasures change, our pleasures can rapidly change into suffering. The events leading to suffering start with a random, subconsciously generated negative thought, then to attachment, to negative emotion, to conscious attachment, to negative feelings, to fixation, and finally to suffering. Continued repetition of this mental routine can lead to an obsession.

Although the history of self-change is dismal, in the case of natural happiness we have a powerful helper—the instructions of our DNA for every cell of our body. In the case of natural happiness, change is self-powered: Each choice will be emotionally rewarded with a flow of joy that, in turn, serves to motivate the next choice. You do not have to renounce your life to enjoy yourself; you can experience happiness in all areas of your life: on the job, socially, standing in line, driving in traffic, living with a spouse, or by yourself.

Expressing spirituality is the easy part. *Spirituality* can be an umbrella term, but this book defines it as a belief in a beneficent God and the practice of altruistic ethics, such as neighborly love and compassion. The reward for spiritual expression is a state of enduring "natural" happiness and joy.

To make a significant behavioral change, it is critical to have a clear concept of the goal, the obstacles, and the strategy. The strategy is to employ epigenetic methods to modify the genes that drive our negative instincts. To ensure the continuance of this procedure, the epigenetic methods must be habitual.

our altruistic genes

The practice of altruistic ethics, such as loving-kindness and compassion, will be familiar to all, but gets frequently lost in the competition for resources and position that comes from our ancient instincts. However, most humans have a moral code, or conscience, that instinctively tells us what is right or wrong, and we are emotionally punished when we violate our conscience. Throughout all cultures, the core differece between right and wrong has to do with selflessness, in the sense of not being self-centered. If behavior is intended to cause suffering to others it is considered wrong; if it brings well-being to others, it is considered right.

Ethics is thought to be about actions and thoughts in relation to others, and the natural effect of causing negative or positive feelings in others is returning the feeling to us. Consequently, the most effective way of bringing natural happiness to ourselves is by acting altruistically, because it creates positive emotions in others, which then come back to the giver. Skeptics may take the position that to do altruistic acts with the expectation that we will be emotionally rewarded is in reality a selfish act. However, our genes see it otherwise. The fact that we know that we will probably receive emotional rewards from altruistic acts is the mechanism by which our genes motivate behavior. It is simply the act of accepting our DNA's instructions, and, by inference, the will of our Designer.

Throughout history, various schools of thought have promoted concepts of ethics based on reason, duty, good, faith, the greatest good, and the like. We go to great lengths to find reasons to promote spiritual ethics, but the reality is that ethics stem from our spiritual genes. Today, ethics have become ever more critical as man has learned to manipulate nature, tinkering with the ecosystem and frequently having disastrous results. Devastation of the rain forest, acid rain, global warming, and myriad other insults to nature cause suffering to many for the gains of a few. Perhaps even more critical are the ethical decisions made in current times with attempts to re-engineer life. With the recent groundbreaking advances in science, politicians, scientists, theologians, and others try to resolve the ethics regarding the latest discoveries in genetics, such as stem cell research, cloning, designer genes, and geriatrics. While our competitive materialism continues to play a part in determining laws and industry standards, our natural sense of morality is pitted against our subconscious negative instincts. Although the moral law of good and evil *is* imprinted on humans, so are the ancient instincts of competition, fear, and violence.

To act in the best interests of another, you have to make a determination relative to the highest good for that other. For example, should you let a 12-year-old drive your car, just because it will make him happy? Though your intention is to give someone else happiness, it must be in his or her best interests. Sometimes you have to consider the difference between pleasure and happiness— giving someone a pleasurable experience is not the same as contributing to that person's sustainable happiness.

pleasure and pain

Much as laugh lines become engraved into the faces of people who often have happy feelings, repetitive negative thoughts and emotions also leave their marks on the brain.

This is the neurological justification for thinking positively.

If you have held the belief that life without suffering equals a happy life, you have no doubt been disappointed. To be free of suffering is no guarantee of happiness, nor do pleasure and suffering automatically exclude one another.

Neuroscientists have identified certain brain chemicals, called neurotransmitters, that are linked to key emotional states. Pleasure results from the production of specific neurotransmitters. For example, dopamine, oxytocine, and beta-endorphin play important roles in desire, contentment, and sexual attraction. Fear, tension, and sadness are associated with acetylcholine and the stress hormone cortisol.

Pleasure is often experienced as a rush. Whether it comes from a massage, a good meal, or sex, the same synapses in the brain are at work. The brain can create endogenous opioid peptides, such as endorphins, enkephalins, and dynorphins. During a peak pleasure moment, you may lose all sense of time and environment—but not for very long; the effect of opioids lasts between a few minutes and a few hours.

Your perception of pain is a neurochemical experience, too. If you cut your finger, pain sensors in your nerves send electric signals through fibers in the spinal cord to the thalamus gland. The hypothalamus can also order the release of opioids to neutralize the sensation of pain by interrupting the transmission of signals from the spinal cord. When a runner is about to be overcome by exhaustion, the release of endorphins and enkephalins allows him to run past the pain so that he experiences a "runner's high." The use of acupuncture for pain relief may be effective because the minor pain induced by the needles results in the release of a very large amount of opioids that override the pain signals in your body.

neutralizing negative emotions

The practice of dissolving negative subconscious thoughts that randomly swirl around in the mind has been practiced for thousands of years in Eastern traditions; the disassociation from our negative thoughts is an important element in Buddhist meditation. Unsurprisingly, becoming aware of the suffering caused by negative thoughts and their emotions is the first step in developing natural happiness habits.

Meditation is a tool for observing thoughts, ideas, and emotions as they pass through the mind. At first, learning to become mindful may feel like an exercise in frustration. Everyone's mind is filled with clutter and "junk" at the beginning. As soon as you begin paying attention, you may find that your mind is like a closet stuffed with all the stuff you did not know what to do with. Just like the objects in a cluttered closet can fall off a shelf and knock you on the head, the "mind-clutter" and "mind-chatter" can overwhelm you when you start to meditate. The most important thing is to set realistic goals for yourself. It is better to plan to give yourself three to five minutes a day of quiet time than to set an unrealistic goal of meditating for 20 minutes to an hour. It takes time to get used to what meditation looks, sounds, and feels like. Like learning to ride a bike, you will fall off a few times and you will feel wobbly and uncertain until you get the hang of it.

Our 28-day program, detailed in Chapter 9, is designed to optimize small windows of time during your day so that you can begin to develop new habits of thinking and observing your emotions. By taking baby steps you can build the skills that will enable you to expand your meditation and mindfulness later on.

cognitive tools

Cognitive psychology offers a range of tools to help you become aware of your thoughts. As you begin to pay attention to the words and ideas that run through your mind unconsciously, you can begin to practice "thought-stopping" by saying to yourself, "Cancel, cancel." Or you can visualize a delete key erasing that phrase from your mind. In the "Reboot" phase of the 28-day program, you will learn a few easy cognitive tools that can help you get control of unhelpful thinking.

preventing the progression of suffering

As you discover how natural it can be to stop negative thinking, you may be surprised to discover that you are suffering less. When you hit the delete key or remind yourself to cancel a negative thought, you are applying the brakes to an internal process that could spiral into a bad mood state. After a few days, your mind-management skills are likely to become positive internal habits.

Without mind-management skills, your instincts can explode into emotions. For example, when you are driving and someone cuts you off, a sudden tide of anger can overwhelm you without warning. Without thought-stopping or mind-management skills, you may react in a way that causes regret later on. But if you take a moment to think, you can interpret the emotion while holding back your instinct to act. You could even put yourself in the other driver's seat and see the situation through his or her eyes. Exercising empathy in this way takes practice, but it yields a profound neurochemical shift in the brain and the body. It may sound strange, and it is not common practice in the United States, but exercising your compassion muscles in periods of crisis can accelerate your happiness makeover.

Channeling a strong emotion such as anger for the greater good is another way to build empathy and compassion for others. Anger *can* be a positive motivator for change—all political and social movements throughout history have been the result of people harnessing their anger against injustice in a way that led to new governments, civil rights, human rights, and freedom for the oppressed. The next time you get angry, ask yourself if there is some way you can use your anger to benefit a community that is suffering injustice. For example, Mothers Against Drunk Driving (MADD) came about because moms who were angry about drunk drivers killing their children chose to organize and educate others.

spiritual addiction

Life experiences show us that nothing is intrinsically all good or all bad; there are no absolutes; and it's a matter of degree and circumstance. Everything in nature has a purpose.

Because spiritual activity causes opioids to flow it can encourage addiction. At first this idea may seem incongruous, but religious extremists have a long history of causing violence and suffering, as you can see on the evening news.

Addiction can be considered a biological accident that happens in the course of the human search for happiness. The same mechanism is at the bottom of all addictions and differs only in the way dopamine and other brain chemicals are released. It is not the search for pleasure that leads people to addiction, but rather the wish to find an escape from life's problems. Addiction develops when a spiritual behavior has embedded itself into the brain as a craving. The experience of addiction permanently transforms the function of the nerve cells in the brain, altering the way in which the genetic information is read and changed into proteins. As do people with other addictions, spiritual addicts have difficult work ahead to overcome their disorder.

the truth of "natural" happiness

It isn't surprising that the religious insights that are most consistent with the recent discoveries of genetics and neuroscience come from the philosophical and religious schools of the East. The difference between the Western faith and Hinduism and Buddhism is their understanding of the source of truth. While Judaism, Christianity, and Islam find truth revealed in a holy book, the religions and spiritual teachers of the East teach us to seek what lies deepest within us.

Though there may be an element of truth in both cases, the message of this book is that the essence of happiness and spirituality lies in our genes. Before the most recent scientific findings provided evidence for this, "truth" was based on subjective experience, observations, opinions, imagination, mysteries, and instinctive beliefs.

Now, you can understand that when you feel joy it is because endorphins and enkephalins are circulating in your head. You may notice an entire fireworks display of flavors when performing a random act of kindness. Everything seems bright and friendly. You feel blessed by a superpower. You feel one with all—beaming perhaps—when you encounter new people. This is not only because of your joy but because these new people actually seem lovable.

conclusion

On a macro scale, our stage of evolutionary consciousness has accelerated due to the scientific breakthroughs of the 20th and early 21st centuries. The deciphering of the genome, genetic engineering, stemcell research, the discovery of a universal field of energy connecting all things, the theory of relativity, and quantum mechanics are some of these scientific milestones.

On a micro scale, as individuals, we have the knowledge to move from a subconscious state to a conscious state using our knowledge of epigenetics. In that transition, it is possible to defuse negative thoughts and emotions, replacing them with positive emotions and acts of compassion, which, in turn, reward us with the happy life that is our biological destiny as humans.

questions for the reader

- What do you think is the purpose of life?
- How is your happiness important to the process of evolution?

- If the purpose of evolution is survival of the species, would not compassion rather than self-serving be more beneficial to human survival?
- If the 28-day happiness plan rings true, why would you not want to practice it?

chapter 9

The Natural Happiness 28-Day Program

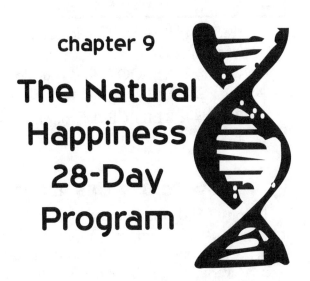

Reminder: *This program is not designed to replace medical or psychiatric treatment for a serious health condition. Please seek professional help if you have questions about your physical or psychological health.*

Knowledge without practice is useless, so this 28-day program is designed to make it easy for you to build natural happiness into your everyday routine. Just as you need to practice your golf swing or a new language, you will see gradual improvement when you use these tools every day.

Common wisdom says that it takes about 28 days—one lunar cycle—to build a new habit.[1] (Addiction-recovery programs sometimes round this out to one month.) With this in mind, your 28-Day Natural Happiness Program is divided into four phases, each of which is one week long. The four phases are:

- Week #1: RELEASE (release stress, worry, and tension).
- Week #2: REBOOT (find your flow).
- Week #3: RELAX (enjoy!).
- Week #4: REJOICE (let your heart and soul awaken).

how to use this program

If you like to plan your work and work your plan, start a journal to record your impressions as you do one or two exercises every day. Set aside a specific time to practice every day. Designate it as your private time when no one can interrupt you for five to 10 minutes. Think of this as your gift or reward to yourself. It costs nothing, and has no monetary value, but the results, as they say, are priceless. A few minutes of soul time can mean the difference between an ordinary day and a special day.

If you do not like to plan and prefer to go with the flow, take a day or two to observe the high and low points of your day and evening. Find one or two windows of time when you can focus your attention on one of the exercises: waiting for a conference call, sitting on hold, at the car wash, sitting in your car between activities, cooling down after a workout, taking a walk, or something similar. Once you start to pay attention, you will discover several pauses during your day when you can focus on a natural happiness exercise.

You have enough stress in your life. This program is designed to relieve some of it, not add to it! Therefore, if you find an exercise is not a good fit for you, you do not have to do it. Move on to the next one or repeat one that you like. Remember: You will get the most benefit from this program *after* the 28 days as you integrate some of these concepts and tools into your daily routine because your mind-body-spirit feels *recharged* and *alive* with continual use.

Be patient with yourself. Be flexible. Have fun! Please let us know how you are doing and if you have any questions, at *www.happinessgenes.com*.

your 28-day program

Week #1: Release

During this first week, you can begin to release stress, frustration, and anxiety from your mind-body-spirit. This eclectic group of exercises combines ancient wisdom (breathing/imagery/mantra) with state-of-the-art cognitive restructuring technology (reprogramming your thoughts to change your mental environment).

DAY #1: THE UNIVERSAL "AAAH!"

30 seconds to 3 minutes

Inhale deeply. Feel your breath leaving your body as you sigh, "Aaah!" The mantra "Aaah!" is a universal sound of release. As you repeat "Aaah!" with each breath you exhale, you can feel tension and stress leaving your body. "Aaah!" focuses your mind on your breath. (For an enhanced "Aaah!" experience, go to *www.templeofsacredsound.org* and visit the Temple of "Aaah!")

What's on Your Not-to-Do-List?

Take an item that is on today's "to-do" list that you do not want to do. Maybe it is stressful, boring, annoying, or you just do not feel like doing it. (Be sure to choose something that is *not* urgent.) Cross this item off your "to-do" list. Create a second list: My Not-to-Do List. Write the item you have crossed off your "to-do" list in your "Not-to-Do" list.

When you are *not* doing that particular activity, please give yourself permission to savor this time you have freed up for yourself. Take a walk. Take a nap. Read something you have wanted to read for a while. Have a cup of tea. Listen to your favorite music. Imagine yourself tomorrow recalling today's Not-to-Do list. Aaah! Feels great, doesn't it?

If you do nothing else, do these two exercises every day for 28 days and they will become an integral component in your new approach to natural happiness.

Days #2, #3, and #4: Color Breathing

30 seconds to 3 minutes

Yawn, stretch, and settle into a comfortable position. Close your eyes and check in with your body: Where are you experiencing fatigue, muscle tightness, or discomfort? Inhale a sense of well-being. Exhale, "Aaah!" Silently, ask yourself, "What color would help my body to feel better?" *There is no right or wrong*; whatever color comes into your mind, let yourself breathe it in and allow it to find its way to any area of your body that would like to feel relaxed and calm. As you release your breath, sense that any tension, fatigue, or discomfort is leaving your body as a different color. NOTE: This is *not* a visualization in which you consciously choose a color. This "spontaneous imagery" process calls upon your creative unconscious mind, which works effortlessly, without conscious effort. You may be surprised at the colors that pop into your mind. You may be surprised that the colors change each time you practice this exercise. *If you are not comfortable with traditional visualization*, you may find it easier to begin by imagining that you can sense a color. Or you may want to imagine that the molecules of a particular color are hitching a ride on the oxygen molecules that you inhale, and flow through your bloodstream to any part of the body that wants to feel relaxed and calm. Remember to exhale a different color: "Aaah!"

DAYS #5 AND #6: TRANQUILITY RITUAL

5 to 10 minutes

Light a candle. Add soothing music—or silence, if you prefer. Watch the candle flame as it changes shape and color. Think: "As my cells regenerate, I enjoy new vitality," and "My cells and my genes function perfectly." If your mind starts to wander away from the candle flame, repeat.

DAY #7: UP, UP, AND AWAY!

7 to 10 minutes

This exercise can be done with a real or an imaginary balloon.

Inflate your balloon. With a marker, write whatever you want to let go of today: worries about money, job, health, family, moving, losing a job, paying bills, a relationship, a friendship, self-esteem issues, weight, and so on.

Go outside and release the balloon into the air. As it gets smaller and farther away, so are those concerns! You can let them go.

Alternately, you can pop the balloon! Doesn't it feel good to zap those issues?

Week #2: Reboot

You have probably seen those television commercials about "Information Overload." Like your computer, your mind-body-spirit gets overloaded from multitasking and being bombarded with too much information and sensory data. When that happens, you need to reboot.

This week's exercises add NLP (neurolinguistic programming) and self-hypnosis to your eclectic natural happiness program.

DAY #1: FIND YOUR FLOW

3 to 5 minutes

Think back to a time when your actions and decisions seemed to flow effortlessly. Maybe you can remember multitasking smoothly while a deadline loomed? Or preparing several courses for a dinner party? Getting your family out the door seamlessly and on time? A day on the ski slopes? Surfing? Playing ball? Pitching a new product or program?

Close your eyes. Ask your mind to take you back to that particular event.

Step into yourself as you were on that day, at that time, so that you can see through your own eyes, hear through your own ears, and feel in your own body whatever you were experiencing at that time.

Make a fist or touch your right or left earlobe.

Experiment with curiosity about how you can maximize the intensity of your flow state by brightening any colors, adjusting the volume of any sounds you associate with this experience, or perhaps slightly shifting your head, neck, or shoulders. Increase the pressure in your hands or on your earlobe as your flow state builds in intensity. When it reaches a peak and begins to fade away like a chord of music, release your fist and your hand. Shake out both your hands.

To reactivate your flow state, make that same fist or touch that same earlobe. Take a deep breath and say to yourself, "Take me back." Your limbic system/emotional brain can replay those molecules of emotion instantly so that you can find your flow whenever you like.

The more you practice this technique, called "anchoring," you will build strength by reinforcing the connection of your fist/earlobe (the "anchor").

DAYS #2, #3, AND #4: COLOR BREATHING AND FIND YOUR FLOW

3 to 5 minutes

Practice color breathing. When you feel fully relaxed, make your fist or touch your earlobe, to find your flow. Say to yourself, "Take me back!"

Combining these two techniques may take you into an altered state of consciousness in which you experience a sense of soul connection to the Universe. Your senses may be heightened to enhance your five senses. Do not be surprised when color, sound, taste, touch, and scent are intensified.

DAYS #5 AND #6: TRANQUILITY RITUAL

5 to 10 minutes

Light a candle. Add soothing music—or silence, if you prefer. Watch the candle flame as it changes shape and color. Think: *My unconscious mind knows how to optimize my genes for health and happiness.* If your mind wanders from the flame, repeat this thought.

DAY #7: PARADOXICAL REBOOT

1 to 2 minutes

This may seem illogical or counterintuitive, but you will be astonished at the results.

Choose a habit that you would like to change. Maybe you would like to stop biting your nails or wiggling your foot during meetings. Perhaps you would like to stop dropping things or being a klutz. You probably have at least one nervous habit you would like to change and you know what that is better than anyone else.

Choose the one you would like to change today. Do it once *deliberately*. *On purpose*.

That's all.

Week #3: Relax

Congratulations! You are halfway through the program.

This week, you are adding a new tool for emotional freedom.

This week, you can reward yourself by taking it easy. We do not want you to develop FDS—Fun Deficiency Syndrome.[2] Please take a day off at the end of this week. It's on us. You have earned it.

Enjoy!

DAY #1: LAUGH!

As much and as often as you want!

Do you have a friend with a great sense of humor? A favorite comedian to whom you can listen as you drive? An old movie? A weekly TV show? You can't help feeling happy when you laugh. In *Anatomy of an Illness,* the late Norman Cousins described how he treated himself to a daily diet of classic comedies after he was diagnosed with an incurable illness. It is worth reading, if only for the miracle ending in which his doctor tells the author that his illness is gone. Apparently, he laughed himself well!

DAY #2: LET YOUR SOUL CATCH UP

10 to 20 minutes

How do you slow down so that your soul can catch up? One Native American tradition was to hike for 10 miles until your body-mind-spirit got into alignment. If you are not quite as athletic or ambitious, take a walk. Spend time in your garden, a nearby park, or a local beach. Feel your feet as they connect with the ground. Native American traditions, Buddhist teachings, and other spiritual practices emphasize the powerful simplicity of connecting with nature in order to let your soul catch up.

Stand still. Wiggle your toes in the earth or sand. Spend a few minutes paying full attention to any sensations—rough or smooth, hot or cold, dry or damp—coming through the soles of your feet. Weather permitting, lie down in a comfortable spot. Relax your body from your head down to your feet and back to your head. Feel yourself letting go of heartache, worry, or anger. Silently give thanks to the soul of the earth and the spirits of the sea and air for receiving and accepting that which no longer serves you.

Days #3 and #4: Emotional Freedom Technique (EFT)

1 to 3 minutes

If you are right-handed, perform this exercise on your right hand; vice-versa if you are left-handed.

What don't you like about yourself? Is it physical? A personality trait? Something else?

Make a series of karate chops against the side of your hand, halfway between the base of your pinky and the top of your wrist. (If you are right-handed, chop *against* your right hand; vice-versa if you are left-handed.) Say aloud three times: "I accept myself even though I am upset."

Now look into the eyes of your reflection. Say aloud three times: "I forgive you for [thing]. I love and accept you as a human being."

Write your thoughts and feelings about this exercise someplace where you can re-read them a few months from now. Keep a pad and pen or a recorder near your bed to keep track of new and interesting dreams that you may have tonight, tomorrow, or perhaps in the next few nights.

DAYS #5 AND #6: TRANQUILITY RITUAL

5 to 10 minutes

Light a candle. Add soothing music or silence, if you prefer. Watch the candle flame as it changes shape and color. Think: *My cells and my genes are in perfect health and balance,* and *I send happy feelings and thoughts to all my genes and cells.* If your mind starts to wander away from the candle flame, repeat.

DAY #7: TAKE THE DAY OFF

Have fun! Your cells and your genes will thank you.

Week #4: Rejoice

During the final week of your 28-day program, you will take time each day to celebrate your new insights, healing, and joy at being alive. You will write a personal statement of reflection or prayer, using an ancient template as a format. Gregg Braden studied the work of spiritual masters from the world's great traditions to discover the hidden template or "lost secret" of creating a message from your soul that can be "read" by the Universe. We invite you to make this reflection or prayer part of your daily routine after you complete this program.

DAYS #1 AND #2: REFLECTION STATEMENT OR PRAYER

20 minutes to 1 hour

"Whether it was conscious or not, the ancients apparently understood how to address the field of energy that connects everything," says Gregg Braden, author of *Secrets of the Lost Mode of Prayer: The Hidden Power of Beauty, Blessing, Wisdom, and Hurt*. "If we can speak to that field in a way that it understands, suddenly, we have the ability to change the genetic blueprint."[3]

This format was known as far back in history as 5,000 years ago. It works!

Any computer program has three basic parts:

- Initialization statements: These bring together everything we need for the program to work.
- Work statements: These do the calculations or get something done.
- Closure or completion statements: These statements bring everything together and wind it up with completion.

When you examine the great prayers of the past, you can break them into these three parts of a computer program. Let's take the Lord's Prayer, for example. It's known as "The Great Prayer." When we really look at how it's constructed, it's amazing. The first statements— "Our Father who art in Heaven, hallowed be thy name"— aren't asking us to do anything. Those are the initialization statements. They're creating an opening of feeling. We're honoring this great presence and creating that feeling in our bodies that opens us to a Presence. We're honoring that name in these "begin" or initialization statements. The second part is the work: "Give us this day our daily bread, forgive us our debts as we forgive our debtors." These are the equivalent of the work commands in any computer program.

The closing or completion statements leave those action statements and make another declaration. We say,

"For thine is the kingdom, the power, the glory forever..."
allahm, amin, in Aramaic, or *amen,* in Christian terms.
This is the closure.

All the world's great prayers follow this format.

Today, spend some time reflecting on what is sacred
to you. Where do you find meaning? What gives your life
depth? Whether or not you believe in God, deities, the
Universe, or a higher power, you can reflect on beauty,
love, appreciation, or forgiveness, or what it feels like to
be cared for unconditionally.

Using Braden's three-step format, write and then read
aloud your personal reflection statement or prayer. Give
yourself plenty of time. Let your heart and soul speak to
you.

DAY #3: JUST BECAUSE

All day

Celebrate your life today. Start by sitting quietly for
a few minutes before you start your day. In your mind's
eye, or in your thoughts, or in your body, rehearse this
beautiful day. See, hear, and feel what it is going to be
like! Spend time doing something special for yourself.
Reach out to a friend who is having a difficult time. Offer
to help someone you do not know. Spend an hour in silence. Read passages from an inspirational book. If you
have a favorite activity—yoga, dancing, paddling a kayak,
taking a drive—do it. Let thankfulness flow abundantly
through your mind-body-spirit today.

Why?

Just because you are...and you can!

DAY #4: FIND OUT WHAT YOUR HEART WANTS

5 to 10 minutes

If you hit the ground running and do not stop until you fall into bed to grab a few hours' sleep before starting again tomorrow, you may feel that your life is busy, yet not full. This can be a sign that you are taking care of what you "should" do at the expense of what you truly want.

Close your eyes. Use color breathing or your reflection statement or prayer to help your mind quiet down. If you find that your thoughts continue to race, imagine that each thought, idea, or comment is a fish floating past you in a giant aquarium. You may be surprised to find that these "mind-fish" slow down as the space between them expands. Don't forget: "Aaah!"

Feel that your attention is a computer cursor. Move the cursor to your heart. Feel it. Listen to it. Observe any spontaneous imagery that may arise. Ask your heart what it wants—what you want. If you are confused by your heart's replies, ask, "What do you *really* want?" Capture your impressions by writing them down. You will want to read them later.

DAYS #5 AND #6: TRANQUILITY RITUAL

Light a candle. Add soothing music—or silence, if you prefer. Watch the candle flame as it changes shape and color. Think: *My joy is helping my genes*, and *I celebrate my choice to evolve consciously*. If your mind starts to wander away from the candle flame, repeat.

DAY #7: PARADOX RITUAL

Today, you can celebrate this elegant paradox: The more you give away what you love, the happier you will be. Today, empty your heart that you may feel it fuller than you ever dreamed!

Trying to grab hold of happiness and keep it for yourself is like trying to catch the wind and put it in your pocket. It will dissolve in your hand. But the more you give away what you love and what makes you happy, the happier you will become.

One of the great psychiatrists of the 20th century, Dr. Milton Erickson, met with a woman who was so depressed she was planning to commit suicide. He asked her what she loved more than anything in the world. When she said her passion was growing violets, he told her to start giving them away—to neighbors, parishioners, students, sick people, or anyone in need. He told her to come back in a month if she still wanted to kill herself.

He never heard from her again. Some 20 years later, he read an obituary in his local newspaper. "The Violet Lady," as she was called, had lived into her 80s and was beloved in the community for the violets that she grew and gave away as gifts. Several hundred people turned out for her memorial service because her unexpected gifts of violets had meant so much to them.

You do not need to be on the edge of despair to give your happiness away. Nor do you need to be a gardener. Tweet your message. Sing. Send an e-card. Write a note. Bake cookies and bring them to work, or send your children to school with cookies for everyone.

Even when the world seems on the edge of catastrophe, or you are living through a tough time, this practice works. Even in unimaginable circumstances such as wartime, there are individuals like a cellist who, during the Bosnian war, played music in the town square. In the background, he could hear machine guns and bombs going off. Yet he continued to play every day, at the same time, in the same place, for years. When he was asked why, the cellist said, "Because the world needs music more than anything."[4] Find your music—the song that wells up when you are living your heart out—and sing it any way you can. We need to hear you.

regroup

If you have had trouble with any of the aspects of this 28-day program, the next chapter will provide additional specific interventions and practices to help to attain—and maintain—good happiness habits.

You do not have to continue this program, although we hope you will incorporate a few of these techniques into your personal, natural happiness tool kit. Feel free to revisit this program at any time.

How has it worked for you? Should we change anything? Replace a few exercises? Move them around? Which exercises could have worked better? Which ones did you dislike? Please let me know. We look forward to hearing from you at *www.happinessgenes.com*.

chapter 10

Internalizing Happiness Habits

The Natural Happiness 28-Day Program in the previous chapter was designed to be a practical, user-friendly, and time-efficient method to build daily habits that will make happiness a natural part of your everyday life. The first eight chapters laid the cognitive foundation for those exercises, as well as the habits that follow in this chapter—as epigenetics research shows, your genes are not necessarily your destiny; you can choose to change your beliefs/ideas, emotions, and behaviors.

Just as the air you breathe and food you eat can tamper with or improve the behavior of specific genes, the beliefs that you reinforce through self-talk, the images that you see through your mind's eye, and your instinctive movements toward or away from people, places, and things can trigger changes in your genes. One of the key differences is that you cannot control a lot of what goes on "out there." If there are work crews digging up your street to install a fiberoptic cable, you cannot change the sound level, but you can alter how you react to the noise.

159

Instead of becoming frustrated, which can trigger a cascade of stress hormones in your system, you can mentally "turn off" the background sound by listening to music or becoming engrossed in your computer. In changing your responses, you are preventing the kind of chemical shifts that can lead to certain genes being "switched on" or "switched off." In the Ornish study of men with prostate cancer, a program of walking, healthy diet, and meditation turned off more than 500 cancer genes, preventing them from being "expressed," or activated. In the study of 12 suicide victims, it was found that the gene that regulates the flow of stress hormones was stunted, so that these individuals were continually flooded with an excess of stress hormones until they could no longer cope. Both studies proved that the internal environment can have a profound impact on how genes behave.

building new habits

Habitual behaviors are automatic. You do not have to think about the first five things that you do when you get to your car, for instance. You learned that sequence when you started to drive. After a few years, it became so automated that you probably cannot tell someone how to do it without a great deal of effort. This is a common example of an everyday habit. It is a routine activity that you do without conscious effort.

As you know, some habits are effective, such as those that enable you to drive safely. Others, such as biting your nails or smoking, are ineffective or unhealthy. One thing all habits have in common is that they are learned behaviors. You were not born doing them!

Learning takes you through four stages:

1. Unconscious incompetence: You do not know how to do X and you are unaware that you do not know how to do X.

2. Conscious incompetence: You recognize that you do not know how to do X but you begin to take the initial steps to learn.

3. Conscious competence: You know what to do but you have to think about it. It does not come automatically.

4. Unconscious competence: You no longer have to think about how to do X because your body/mind has absorbed and integrated the steps into one fluid process. In fact, if you had to explain how you do X, it would take effort for you to remember.

Cognitive habits (sometimes called "internal habits" or "mental habits") are beliefs that you reinforce through repetition, called "self-talk." Emotional responses that recur through time can cause you to create a set of beliefs about yourself or about life. For example, if you are afraid of elevators and you get butterflies in your stomach whenever you picture an elevator in your mind's eye, your mind may explain that to you with a belief that states, "Elevators are dangerous. Do not go near them. Ever!" Perhaps you got stuck in an elevator when you were a child or you saw a movie about people stuck in an elevator that frightened you. Whatever the cause, your emotional and physical reactions (called "kinesthetic" reactions) override your logical mind.

your subconscious mind
drives your personality

According to Freud, the psyche has two parts: the conscious and the subconscious. The conscious mind is the logical thinker. When you think, "This is who I am," you tend to think of your conscious mind as "the self." But the psyche is like an iceberg. The conscious mind is like the tip of the iceberg. It is the part that is visible but it represents only a small fraction of the whole. The majority of the psyche, like the larger part of the iceberg, is beneath the surface! In the case of the iceberg, the bulk is submerged under the water. In the case of your mind, the majority is submerged below your level of attention. This is how it came to be named the sub- (under) conscious mind (also termed the *unconscious mind*).

The conscious mind says, "I should not be afraid of that elevator."

The subconscious mind yanks the emergency cord and brings everything to a halt. Inside, a part of you is shrieking, "Yikes! No elevator for me!"

Which part do you listen to?

That's right. The subconscious mind has a lot more power. No matter what your logical mind says, if your subconscious mind wants to hit the brakes—or drive the car—you will find yourself doing what it wants.

When it comes to your happiness, any number of stressors and underlying themes have probably become habitual ways of thinking and feeling. If you lost your job and are having a hard time finding a new one, you may remember times in the past when things did not go your way. Before you realize it, you are thinking, "Nothing turns out the way I want it to." It does not take long before that belief is followed by sadness and defeat. Soon, you start to believe, "I will never be happy."

Believe it or not, these thoughts that are repeated silently in your mind are habits that can be as toxic as smoking three packs a day. Not only do such "cognitive habits" prevent you from feeling upbeat or curious about new possibilities, but they also increase the level of stress hormones and make it more difficult for your brain to keep producing healthy neurotransmitters. As you continue to reinforce this damaging self-talk, the increased stress can lower your immune system and prevent those genes that are essential to your happiness from being activated.

The good news is that because a habit is something learned, it can be unlearned. You need to find it, identify it, and practice a few mind-changing exercises, but yes, you can change it. Those beliefs and emotions that are toxic to your well-being can be released, and you can reboot your mindset!

Remember this old riddle?

Q: How do you get to play music in Carnegie Hall?
A: Practice, practice, practice.

hypnosis and self-hypnosis

Hypnosis is a form of a mind/body intervention that goes back thousands of years. The term *hypnosis* comes from the ancient Greeks, whose god of sleep was named Hypnos.

Although hypnosis is often considered a parlor trick, there are distinctions between hypnosis for entertainment and hypnosis for therapy. In both settings, hypnosis is a method of producing a deep state of relaxation such that our subconscious is more receptive to suggestions. A stage hypnotist makes suggestions that are harmless to the participant yet funny to the audience. The intention of both performer and volunteer is to entertain the audience. On the other hand, therapeutic hypnosis takes place in a one-on-one context in which the objective is to help a subject draw upon the healing resources of his or her subconscious mind for the purpose of healing a belief, emotional pattern, or behavior.

There are many levels of the hypnotic state; for instance, you may experience driving down a highway when all of a sudden you become aware that you have missed your turn. You were not driving irrationally, and yet your mind was focused on something else. You were actually in a hypnotic state. Other examples include daydreaming, reading a book, or watching TV. Even words can be hypnotic, and they can stimulate your imagination, creating a suggestive state in your mind. Meditation and prayer are also forms of a hypnotic state. In other words, whenever you are doing something you are not consciously thinking about, you are in a state of hypnosis.

Self-hypnosis is, as it sounds, a self-induced form of hypnosis, which you can apply to virtually any activity or outcome that depends on your own efforts, or that you can influence with your mind. Sometimes it is so easy the results seem mysterious and even miraculous, but it works in a consistent manner that can be explained, predicted, and repeated. It's biology, not magic. You are simply learning how to focus the conscious mind on relaxing deeply and easily while you suggest to your subconscious mind that it can enjoy new levels of natural happiness.

With sufficient practice, you will be able to give your-self a posthypnotic suggestion that will prolong your de-sired results after your self-hypnosis session has ended. The use of posthypnotic suggestions can be very power-ful. When you use only your willpower to change a habit or behavior, you are only using about 10 percent of your mind, but when you program your subconscious mind to change your whole mind, and keep the change, you are tapping into the hidden power of that 90 percent of your mind that is submerged beneath the threshold of your attention.

The subconscious mind does not philosophize or think logically. It will only accept what it directly expe-rienced, which can also include that which is conscious-ly imagined. If we tell you to imagine a nice, juicy lem-on and hold it in your hand...now cut a slice and put it in your mouth...what happens? Doesn't your mouth salivate? Your saliva glands are part of your autonomic nervous system, meaning that they are outside your con-scious control. Try telling yourself to salivate and your mouth may even go dry. But when the autonomic ner-vous system is presented with a powerful suggestion, such as the idea of a lemon, you just cannot help yourself from salivating. Even though there was no physical lem-on in your mouth, your unconscious mind believed that the lemon was real, and within a nanosecond, triggered your saliva glands.

That, in a nutshell—actually a lemon skin—is the heart of hypnosis. It is as direct and simple as imagining something in a way that is so compelling that your body believes it is real! All the while, you feel comfortable and relaxed.

elements of happiness habits

goals

The first step is to formulate a specific definition of exactly what you want to accomplish. Saying you want to be naturally happy may make conscious sense, but such simple goal formulations won't get you very far with the subconscious mind. To have any influence with the subconscious mind, goals must be clear and unambiguous. When you are formulating your suggestions it is necessary to know the specific behaviors you must develop to achieve your goal.

suggestions

The power of suggestion has substantial influence on our behavior, beliefs, attitudes, and values. We are wired to continually monitor information coming to us through the five senses, always on the lookout for opportunity, and, especially at the subconscious level, on guard against threat. Much of the information that is important to us is suggestive in nature. In short, we are built to respond to suggestion. The suggestions we are talking about don't promote a voluntary response, but rather a hypnotic suggestion that will produce a non-voluntary response. The effectiveness of suggestion has been demonstrated over and over again in every field of medicine and human behavior. A prime example is the use of placebos. To get the placebo effect, one must think that he or she is being treated with a medicine. Advertising is another excellent example of the practical application of suggestion. Most people say that advertising does not affect them, but it is working if sales go up when advertising is increased. To illustrate how suggestion plays a role in almost all areas of behavior, consider the commonly used examples of Pavlov's dogs. When Pavlov rang a bell that he had previously associated with food, the dogs would salivate. He had discovered classical conditioning, or is a conditioned response to suggestion. We all have our own equivalent of the Pavlov's bell example.

In self-hypnosis, suggestions of one or more goals to the subconscious mind heighten its receptivity to new habits. But while we are sensitive to the power of suggestion, the primary function of our subconscious is to protect us, and consequently will reject suggestions that it perceives to be threatening or not in your best interest. For example, if you subconsciously believe that meat is necessary for better health, then a suggestion to change your eating preference to plant foods will not be effective.

affirmations

Suggestions include affirmations, and can be verbal, silent, or an image. Depending upon whether your mind prefers visual images or words, you will respond to either visualization or verbal suggestions. Every person is different, and not everyone can visualize comfortably. Experiment with yourself to determine whether imagery or language appeals to you more. Remember: There is no right or wrong! You want to find the best method for yourself. By "best," we mean the most comfortable. Do not force yourself to visualize if you respond better to language. Nor should you blame yourself if your mind prefers visualization. Whatever the format, when you relax and add suggestions for being happy, your subconscious mind will follow through by creating new habits that support that belief. Although this sounds oversimplified, hypnosis can install the belief that you are happy, which, in turn, makes happiness more likely. Work with the subconscious must be done in a state of relaxation rather than under stressful conditions. You can reinforce self-hypnosis with a smile! That is a posthypnotic suggestion anyone can remember.

posthypnotic suggestions

The most powerful tool in self-hypnosis is the posthypnotic suggestion. When in a hypnotic state, the conscious mind loses its ability to make critical decisions, and during that time your subconscious can accept a suggestion as reality. A powerful suggestion continues to function after the hypnotic state has ended, and your conscious mind accepts your suggestion as part of its decision-making process. The deeper the hypnotic state, the easier it is to implant the suggestion into your subconscious mind. The more you experience the suggestion as a reality, the more it will actually be a reality in your waking state.

posthypnotic triggers

A trigger produces a reaction to a posthypnotic suggestion that was planted in your subconscious mind during a self-hypnotic state. It may be a word or a touch, a sound or a visual cue, a smell or a taste, or an emotion. The stronger the suggestion you implanted, the greater effect the trigger has in re-creating the response. You can then create a posthypnotic suggestion to respond to a positive trigger that you implanted in your own subconscious mind.

When you are practicing a hypnotic induction (explained in detail in Appendix II), and you count down to zero, tell yourself that you will feel very comfortable in a favorite place in your mind that is calm and relaxing. This place may be from your memory, a new place, or just a place deep inside yourself where you can have positive experiences. Beautiful places in nature, such as gardens and ocean beaches have wide appeal. Once you have entered a hypnotic state, tell yourself to place your thumb and forefinger together (or use your preferred word or a touch that is meaningful to you). This will be your trigger, and it will remind you of how relaxed and calm you were feeling at that moment. Every time you practice self-hypnosis and reinforce your trigger it will become stronger and stronger in your subconscious mind. Each time you trigger your suggestion it will work better and be more responsive.

If you do not see improvement by the end of three weeks, we suggest that you consult a professional hypnotherapist or a psychologist trained in hypnosis. Sometimes you are too close to the problem and cannot see the forest for the trees. It can be challenging to develop self-hypnosis skills from a book because you have to re-read the steps and repeat them while closing your eyes. Success may come more easily when someone is giving you verbal suggestions and all you have to do is follow his or her voice. The best results are likely to come when suggestions feel natural and effortless. Another method is to follow your own voice, by recording the script and playing it back. Bedtime is an excellent time to apply your suggestions. Just before going to sleep, repeat your verbal suggestions or visualize your image suggestions. This has a way of loading them into the subconscious mind, which will work with them while you are asleep.

mind-body interventions and epigenetics

Until recent research in genetics, most scientists thought that the human body was a biochemical machine programmed by its genes, and that our abilities (artistic and intellectual) and our health (diseases, depressions, and so on) were genetically programmed. But now, stimulated by the Human Genome Project and the new science of epigenetics, it is recognized that our environment, and more specifically, our perception (interpretation) of the environment, influences the expression of our genes, through a process known as *epigenetic energy*. This new perspective of human biology incorporates the role of a mind and spirit in all healing, for when we change our perception or beliefs we send totally different messages to our cells and reprogram their expression. This may well explain why some people seem to have had spontaneous remissions or recovered from disabilities thought to be permanent.

developing spiritual habits

As described in previous chapters, popular self-help programs attempt to sell the concept that by using willpower we can override unhealthy habits and adapt healthy ones. But this is not the case. If you could use willpower, you would need to keep a constant vigil on your habits, because the moment you lapse in consciousness, the subconscious mind will automatically play its old habit program. The subconscious mind can be thought of as a tape player that records from genetic and environmental input without any supervision from the conscious mind.

While the conscious mind does the thinking, creative, and decision-making part, the subconscious mind is like a supercomputer loaded with a database of programmed behaviors. Some of these programs result from genetics, such as instincts; however, others result from nurture and beliefs from our learning experiences. The subconscious mind is not used for reasoning or creative consciousness; it strictly responds to a stimulus. When an environmental signal is perceived, the subconscious mind reacts with a stored behavioral response—no thinking is involved.

Consequently there is no discernment of whether the subconscious behavioral program is good or bad. Furthermore, the subconscious mind has much greater power than the conscious mind, as an information processor. Neuroscientists state that the conscious mind only provides about 5 percent, whereas the subconscious provides about 95 percent, of our daily cognitive activity. So, no matter how determined you are to change your unhealthy instincts and habit tapes in your subconscious, you have to rerecord the program with the desired new habits.

There are essentially two ways to change old habits. First, we can become more conscious, and rely less on automated subconscious programs. By being fully conscious, we have more control over our ability to develop new habits.

stress and your health

The 28-day program in Chapter 9 produces such healthy side effects as increasing spirituality, improving your baseline mood state, and creating a consistently optimistic mindset. Research shows that you are also shifting the expression of genes that are integral to your happiness. Your mind and your physiology flow together on a continuous basis for greater joy.

Stress is how the body reacts to change. As defined by Dr. Hans Selye in the 1920s, stress is neither good nor bad. Eu-stress means health stress, such as getting a raise, celebrating your birthday, or moving to your dream home. Dis-stress means unhealthy stress. Dis-stress can lead to lowering of the immune system, a rise in blood pressure, an increase in headaches, and a host of gastro-intestinal disorders. Some people say that as much as 85 percent of visits to physicians' offices in the United States are due to dis-tress.

The stress hormone cortisol has the same chemical precursors as DHEA, is our most common hormone, which is associated with many vital health functions, such as longevity and cell repair. Both cortisol and DHEA stem from the adrenal glands, so when the adrenals are manufacturing the higher-priority survival hormone cortisol, DHEA production falls off and our health is negatively impacted.

In addition, the body's stress response initiates a vast array of chemical reactions, and causes many hormones and neurotransmitters to shift in response to stressful stimuli. Consequently, stress management is vital for happiness and health. Virtually all of the modalities of this program for Happiness Habits, such as mindfulness and the relaxation response, not to mention love and compassion, can be considered natural de-stressors. In effect, it can be said that we are epigenetically engineering our own cells for natural happiness and health.

laying the groundwork

What follows is a suggested schedule of exercises/interventions that will prepare the subconscious mind for accelerated results. Just as a gardener prepares the soil prior to planting seeds, these exercises help to condition the subconscious mind to be receptive to new ideas, emotional patterns, and behaviors that will lead to new habits of natural, sustainable happiness.

program outline

THE FIRST TWO WEEKS (PRACTICE EACH DAILY)

- Act of mindfulness.
- Act of compassion.
- Act of agape love.
- Act of relaxation response.

THE THIRD WEEK

- Perform the four acts twice daily.
- Practice self-hypnosis for 10 minutes a day.
- Practice verbal suggestions for 5 minutes a day.

THE FOURTH WEEK

- Practice an altruistic act four times weekly.
- Practice self-hypnosis 15 minutes daily.
- Practice verbal suggestions daily.
- Practice posthypnotic suggestions weekly.
- Practice affirmations four times weekly.

mindfulness

The practice of mindfulness will help you to become aware of mental habits buried in your subconscious mind. This mind/body technique is frequently referred to as "living in the moment." In order to make any change, you must first be aware of what you are thinking and feeling. Throughout the program, the practice of mindfulness is critical to your developing mastery over the hidden beliefs and automatic patterns that may be holding you back.

Mindfulness is a natural mental ability that is available to anyone, regardless of culture, ethnicity, or background. It is not a belief or tradition, religious or scientific, but rather a practical, natural way to be aware of your life, moment by moment. The practice of mindfulness is nothing more than being non-judgmentally aware of your thoughts, feelings, and environment in the present moment. From a viewpoint of Eastern consciousness, the Western state of consciousness seems like an extended dream that prevents you from waking up to experiencing the fullness of life. With daily practice of mindfulness, you can wake up from this "dream life" and participate in real life as it unfolds moment by moment. With the power of mindfulness, you become aware of your subconscious beliefs and instincts so that you can focus your awareness. This state of "right concentration" is the state in which world records are set, visionary concepts are developed, and you are at your peak efficiency. One goal of practicing mindfulness is to clear your mind of extraneous ideas and emotions so that you can focus on the epigenetic exercises presented in the 28-day program.

When you are not living in the present moment, your thoughts are in the past or the future. You may think this is a way to run away from your problems, but the opposite is true: Mindfulness helps you to see your problems more clearly so that you can resolve them. Just think of the most creative and productive people whom you respect. One element they share is an ability to block out distractions and focus completely on the tasks at hand.

Mindfulness should be practiced several times a day so that the random thoughts bouncing around your mind start to calm down. This is a great technique for energizing yourself and creating sustainable levels of natural happiness.

One excuse Americans often use to justify their refusal to practice mindfulness is that they do not have enough time. However, in reality, practicing mindfulness a few minutes a day will free up your mind to become more productive for more hours per day than before you practiced mindfulness. In fact, mindfulness can increase your efficiency. As with self-hypnosis or any of the other exercises offered in the 28-day program, you must practice every day.

Do not be surprised when you find that you have greater energy available to you toward the end of the day.

This is because a daily practice of mindfulness helps to clear the mind of unproductive ideas and negative emotions that sap energy. You may well discover that after about 10 days of practicing mindfulness you function more effectively because you are no longer wasting energy due to anxiety, fantasy, or smoldering resentment. At the end of a full day's work, your mind retains its clarity and focus.

the basics of mindfulness

Your breath is your link to a state of mindfulness. As you sit and focus on your breathing, you will become aware that your mind drifts away from your breathing very easily at the beginning. When you find yourself thinking of anything but your breath, simply bring your focus back to your breathing. The longer you can pay attention to your breathing, the more centered and focused your mind will become.

Breath is the metronome of our life, and it responds to balance changes in bodily states. It sets the rhythm for our body, and is always there. Because your breath is always with you, you can practice mindfulness in any location: in the privacy of your home, in a natural setting, or even waiting in line. Breath is the bridge to your mind.

Breathe normally for a few cycles of breath. Then extend the exhalation cycle so that you are emptying your lungs of more air. The next breath that you inhale will become that much stronger. Pay attention to each breath for its full duration, the entire inhalation, and the entire exhalation. Do not try to control your breath—just be aware of it. If your breathing gets shallower, let it be shallow; if it gets faster or slower, let it. The breath regulates itself. Do not try to force it or change it, but just feel it and be aware of it.

the practice of compassion

Ranked a great virtue in numerous philosophies, compassion is also considered in all the major religious traditions as among the greatest of virtues. The Christian Bible's Second Epistle to the Corinthians is just one place where God is spoken of as the "Father of compassion" and the "God of all comfort."

The life of Jesus embodies for Christians the very essence of compassion. Christ's example challenges Christians to forsake their own desires and to act compassionately toward others. In the Parable of the Good Samaritan he holds up to his followers the ideal of compassionate conduct.

More vigorous than empathy, compassion motivates an active desire to alleviate another's suffering. It is often, though not inevitably, the key component in what manifests in the social context as altruism. In ethical terms, the various expressions down the ages of the so-called Golden Rule embody the principle of compassion: *Do to others what you would have them do to you.*

On a practical basis, compassion can become a conscious act that can bring immediate natural happiness into your life. The key to developing compassion in your life is to make it a habit.

After you have practiced an act of compassion once a day and have experienced the emotional rewards, you will find more opportunities. In addition to finding that acts of compassion bring you happiness, you can refer to some of the scientific studies in this book that show people who practice compassion, produce substantially more DHEA, the hormone that counteracts the aging process, and much less cortisol, a stress hormone.

COMPASSION PRACTICES

Empathy Practice. The first step in cultivating compassion is to develop empathy for other people. Try this practice: Imagine that a loved one is physically or mentally suffering. Now try to imagine the pain he or she is going through; see, hear, and feel it with all your senses. After doing this practice becomes easy, try extending it the suffering of others who are not close to you.

Oneness practice. Instead of looking at the differences between yourself and others, think of the similarities. Although your external appearance is different, inside yourself, your needs and desires are shared by billions of other human beings.

Affirmations. Choose one of the following statements and apply it to your life. Repeat this phrase at least five times per day for one week.

- "Just like me, this person is seeking happiness."
- "Just like me, this person is trying to avoid suffering."
- "Just like me, this person has known sadness, loneliness, and despair."
- "Just like me, this person is seeking to fill his or her needs."
- "Just like me, this person desires to be compassionate."

Relief of suffering practice. Once you can empathize with another person, and understand his humanity and suffering, the next step is to want that person to be free from suffering. This is the heart of compassion. Reflect on how happy you would be if another human being desired your suffering to end, and acted upon it.

Act of kindness practice. Choose one small behavior every day that is aimed at alleviating another person's suffering: a smile, a kind word, doing an errand for someone else, or listening. Make this a daily practice. Eventually, it will inform your choices throughout the day, every day.

Those who mistreat us practice. The final stage in these compassion practices is to want to ease the suffering of not only those we love, but even those who mistreat us. Reflect on such a person. Is it your boss, a stranger who cut you off in traffic, or a family member? Did someone blame you for something that was not your fault? Whatever the event, spend a few minutes being curious about that person's personal history. What was he or she taught as a child? Try to imagine the mood and state of mind that person was in, the suffering that person must have been going through to mistreat you that way. Acknowledge that he or she was doing his or her best, just as you were when you may have mistreated someone else because you were having a tough day or a miserable week. Compassion and forgiveness are necessary ingredients for natural happiness.

agape love

Today, the word *love* is used to convey a broad list of subjective meanings, such as *I love my dog, I love my car, I love my husband, I love mashed potatoes,* and endlessly on. But in the days when the New Testament was written in Ancient Greek, the Greek language had four distinct words for love: *agape, eros, philia,* and *storgē. Agape* is the main word used for love in the New Testament. The term *agape* was used by the early Christians to refer to the self-sacrificing love of God for humanity, and it is the type of love taught by Jesus. When He says "God is love," the Greek New Testament uses the word *agapao* to describe God's love. The other Greek words for types of love are: *Eros,* a romantic, or lustful "Hollywood" type of love, which has the most pleasures, depending on the situation and circumstances. Eros is not used in the New Testament. *Philia* is the love of friendship, best friends, and the fellowship of being with those people you enjoy. Philia love can be pleasant, but not reliable, as it's also held captive by the sifting sands of situation as well as by our and others' perceptions and expectations. Unfortunately, we probably all know of a friendship that waned or was severed because of time, distance, harsh words, or the way someone interpreted another's actions. *Storgē* love is love of family, which has a genetic link for survival. Although it can be a provider of protection, as the well-known phrase "blood is thicker than water" suggests, it can be broken by competition and opposing interests.

Unlike the previous three types of love, agape is not held hostage by its environment or someone's perception. The reason agape can soar above these things is that it is based upon the commitment of a decision. It entails the decision to proactively seek someone's well-being. Because it is not a knee-jerk reaction or just a responsive feeling to how you've been treated, agape is capable of acting in a hostile environment where there are no warm fuzzy feelings. For example, Jesus' teaching that we should *agape* our enemies is intended to show the boundless nature of the Christian commitment toward seeking another's well-being (Luke 6:35).

The New Testament is full of examples and teachings illustrating the nature of agape as well as teachings designed to train the disciple's heart to be shaped by agape. A few examples illustrating the active nature of agape include:

- "Knowing that sinful man would kill His Son, but also knowing that without Jesus we were doomed, God loved [agape] us by sending his Son." (John 3:16)
- "Those who love [agape] Jesus will do what Jesus taught." (John 14:15, 23)
- "If a person has material resources and the love [agape] of God within him, his heart will take care of his brother who is in need." (1 John 3:17)

As if it were not enough that the proactive nature of agape has the power to rise above its environment, it can also empower passion and friendship! For example, when a spouse chooses to speak and act toward his or her mate with agape, this creates the loving environment in which eros and philia can thrive! Although the proactive spouse might even perceive the other spouse as being unkind or rude, additional problems can be prevented by responding out of agape while the power of agape works at nurturing the growth of the other forms of love! Jesus taught his disciples that the world would know that they were his people if they would show agape toward one another. (John 13:35.)

relaxation response

The most critical step for communicating with your subconscious is to enter a state of relaxation. The Relaxation Response, as taught by Herbert Benson, MD, at the Benson-Harvey Mind-Body Medical Institute at Massachusetts General Hospital in Boston, can be understood and applied by anyone, anywhere. The National Institutes of Health have studied it and recommend it as effective for stress reduction.

Dr. Benson and his team of researchers from Harvard have conducted thousands of studies on the power of meditation and the Relaxation Response throughout the course of more than 30 years. With the permission of His Holiness the Dalai Lama, Dr. Benson studied a group of Tibetan monks wrapped in towels soaked in ice water prior to a meditation session in a cave where the temperature was measured to be around 20 degrees Fahrenheit. In a landmark film, the monks are seen meditating until the towels begin to steam from the heat they generate inside their bodies. If there was ever any doubt that the mind can produce physical changes, this pilot study put it to rest. (The film is available on DVD at Massachusetts General Hospital's gift store.)

The following is Dr. Benson's relaxation technique, in a nutshell.

Take a few deep breaths. Close your eyes. Select a word that conveys relaxation and peace to you. Some people choose a phrase from a prayer; others prefer a word such as relax.

Repeat it slowly as you continue to inhale and exhale. Relax as you breathe. (That's right: It's so simple, even a caveman can do it!) When you open your eyes and stretch, you will feel as relaxed as if you have had a long, comfortable nap. Your mind, body, and spirit will feel refreshed.

AUTOGENIC RELAXATION

Sometimes called *progressive relaxation*, this process is used to introduce the body and mind to a full experience of relaxation.

To experience this, while tensing and releasing each area of your body, repeat the word relax. *Make yourself as comfortable as possible. You may sit or lie down. Take a breath of air, filling your lungs to a comfortable level, and focus on an object in front of you. Exhale slowly. Then fill your lungs again. Slowly let your eyes go out of focus as you continue breathing in this slow manner. From the feet to the head, start relaxing the muscles throughout your entire body in an orderly sequence. First, feel your feet relaxing, and then feel the tension going out of the muscles in your lower legs. Continue to relax the muscles in your lower body. Move upward to your chest, arms, and neck. It normally takes between 10 and 20 minutes to relax completely.*

a few final thoughts

These mind/body exercises have been researched and found to be effective. You will find some that work perfectly for you on the first trial and others that seem awkward. Do not force yourself to do exercises that are not enjoyable. Remember: Nothing works all the time for everyone—not even aspirin. Certain mind-body interventions will appeal to certain people, whereas those same techniques will frustrate others. If you practice the exercises in this book mindfully, chances are excellent that they will lead to your developing new habits of happy thinking, feeling, and behaving.

Although these practices and the 28-day program are designed to deliver noticeable results in about a month, it may take you longer to observe improvement. Or, you may find that from the very first day, your mind-body-spirit respond immediately. There is no time restriction, and we ask that you be patient and experiment with different combinations of practices to find which ones work best for you. We would appreciate your feedback so we can tweak this program. Please let me know what has worked really well for you and what can work even better. We need your help to keep improving this program. You can reach me at info@happinessgenes.com or Tweet me @happinessgenes.

afterword

Welcome to the Revolution!

If you have read this far (or perhaps turned here to read the ending first), you may not realize it yet, but you are now part of a silent revolution.

That's right.

Not to worry. Nothing is going to die, except some old ideas that no longer hold true.

We are talking about a revolution in evolution. Biological, spiritual, and intellectual evolution. *Evolution* means a process of gradual change. *Revolution* signifies a quick turnaround from the status quo to a new way of living.

In the past five years, a stunning amount of new findings have accelerated humanity's access to information that is only now being discovered. With that information comes an array of opportunities to expand our range of knowledge today. As we find new ways to apply this new information, we can assist future generations as well. Never before in human history has this portal for conscious evolution been open to us.

My thanks go to the scientists who are leading the revolution in evolution. But you, the reader, have an important role to play as well:

Question answers. For your personal growth and intellectual evolution, questioning answers is far more important than answering questions.

Open your mind. Your mind is like a parachute; it only works when it's open. Corny, perhaps, but true. You do not have to believe or disbelieve anything about this controversial subject in order to benefit from its research and emerging concepts. You do, however, need to keep your mind open to the possibility that there is something here that can help you lead a happier, healthier life.

Seek truth. Whether you approached this book as a disciple of science or a spiritual seeker is irrelevant. The spirit of science and the science of spirituality seek truth. Genius is the ability to hold two seemingly contradictory concepts in your mind without eliminating one of them in order to be "right."

We sincerely hope we have answered at least some of your concerns. We recognize that epigenetics is a complicated subject and you may find yourself confused after reading this book. Although uncomfortable, confusion may motivate you to continue seeking information. At the end of the day, you may discover that you are confused on higher level and about more important things.

Behavioral Science and Epigenetics in Depth

This book has been concerned with the elements of behavioral genetics and spirituality.

The purpose of our research was to link our spiritual genes with natural happiness, and in the process we uncovered the genetic and environmental obstacles that block natural happiness and allow suffering and violence. In order to help you remove these obstacles on your journey to natural happiness, we have presented a program of epigenetic methods in Chapter 9.

Because the elements involved are not commonly known, the purpose of this appendix is to provide a more in-depth understanding, for those interested.

genetics

Genetics is all about physical traits and the code carried in DNA that supplies the building plans for any organism. It is impossible to overestimate the influence of genetics on our planet. Every living thing depends on

DNA for its life, and all living things, including humans, share DNA sequences. The amazing similarities between our DNA and the DNA of other living things suggest that all living things trace their history back to a single source. An organism's genes control how it looks, functions, and reproduces. Because all biology depends on genes, genetics is a foundation for many other sciences, including agriculture and medicine.

The gene is a basic unit of heredity found in the cells of all living organisms, from bacteria to humans. They are composed of segments of DNA, a molecule that forms long, thread-like structures called *chromosomes*. The information encoded within the DNA structure of a gene directs the manufacture of proteins, molecular workhorses that carry out all life-supporting activities within a cell. *Gene expression* is a term relating to the fact that our genes are not active all the time; because every cell in our body carries our entire set of genetic instructions, only certain genes are active at the same time and working for each tissue type. The genes that might interfere are then turned off (not expressed), like a light when you leave the room. Gene expression is controlled by hormones, which circulate in the bloodstream and affect the appropriate tissues. Consequently, hormones act like a master switch for gene regulation all over the body.

In organisms that use sexual reproduction, offspring inherit half of their genes from each parent and then mix the two sets of genes together. This produces new combinations of genes, so that each individual is unique but still possesses the same genes as its parents. As a result, sexual reproduction ensures that the basic characteristics of a particular species remain largely the same for generations.

However, mutations, or alterations in DNA, occur constantly. They create variations in the genes that are inherited. Some mutations may be neutral, or silent, and do not affect the function of a protein.

A mutation may benefit or harm an organism, and throughout the course of evolutionary time, these mutations serve the crucial role of providing organisms with previously nonexistent proteins. In this way, mutations are a driving force behind genetic diversity and the rise of new or more competitive species that are better able to adapt to changes such as climate variations, depletion of food sources, the emergence of new types of disease, and changes in beliefs and behavior. Research has confirmed that the human genome is still evolving: Some genes get duplicated and have new functions while others lose their function. The dynamic nature of the human genome suggests that our genes, likewise, are evolving

heredity

Presently there is a lack of real evidence for the heritability of human behaviors and personality traits. Although there are many studies, opinions, and other material on this subject, the best evidence appears to be in the study of identical twins. If you had personal contact with a set of identical twins you would see that they share a remarkable physical resemblance as well as certain mannerisms, but their personalities are not necessarily the same. Throughout many years, scientists have studied identical twins in order to determine the influence of nature and nurture. Such studies also included analysis on identical twins who were adopted to different homes at birth. A recent study of genetic happiness derived from identical twins conducted by David Lykken, disclosed in his book *Happiness: What Studies on Twins Show Us About Nature, Nurture, and the Happiness Set Point,* indicates that some 60 percent of identical twins separated at birth are likely to describe themselves as happy.

Future genetics studies may well provide genetic predispositions for certain human behaviors. For instance, common variants in a receptor for the neurotransmitter dopamine are associated with an individual's ranking on a "novelty-seeking" trait, and a variant in a transporter for serotonin is associated with anxiety.

One of the most controversial traits of human individuality is that of intelligence, because there is much disagreement about the definition of intelligence and how to measure it. This emotionally charged subject is increasing in interest because there appears to be a strong heritable component involved. At present, no specific DNA variant has been demonstrated to influence IQ, and it is generally accepted by researchers that many related variants will be uncovered.

The possibility of genetic spirituality is currently a hot topic, and has been further fueled by immense media coverage of the publication of a book called *The God Gene*. The author/researcher of this book, Dean Hamer, used personality testing to indicate that a trait called "self-transcendence" showed heritability in families and twins. This trait was associated with believing in something that can't be proven, and was demonstrated by experienced mediators. Because personality traits usually have some heritable influence, this finding would not be unusual, but Hamer claimed that a variant gene, VMAT2, was associated with a high score on the self-transcendence scale. Francis Collins, director of the Human Genome Project, in his book *Language of God* went on to state that, as none of Hamer's data have been peer reviewed or published in the scientific literature, most experts reacted to his book with considerable skepticism.

According to Collins, there is an inescapable component of heritability in many human behavioral traits, but environment, and particularly childhood experiences and free will, have a profound effect on us. Future scientists will discover an increasing level of molecular details about the inherited factors that underlie our personalities, but that should not lead us to overestimate their quantitative contributions.

Although this seems to be the current mainstream scientific position on the subject of genetics and spirituality, there is convincing evidence that, throughout history, humans have had a spiritual nature and an unquenchable hunger for God. In fact, the major goal of Collins's book is to present evidence that God exists. To him, it is self-evident that we have a longing for God, and receive rewarding emotions for spiritual acts such as doing a good deed, or for an act of compassion. Waiting for the discovery of specific genes that may contribute to our spirituality may be the course for some scientists, but most prefer to accept natural happiness now.

human behavior

Behavior refers to the actions or reactions of an organism, usually in relation to the environment. Behavior can be conscious or subconscious, overt or covert, and voluntary or involuntary. Human behavior can be influenced by a large number of circumstances, such as culture, emotions, ethics, authority, hypnosis, coercion, and genetics. Predicting behavior is far more complex than is commonly thought. Successfully predicting or changing behavior has a long record of failure. Shakespeare has tried to describe it, and Freud, Jung, and other psychiatrists have sought to explain it, but all failed. Today, personal columnists offer their advice on how to correct, predict, or change the behavior of another or oneself. Countless books on self-help tell how to do it. Weight-loss programs teach changing eating behavior. Spouses relentlessly try to change the behavior of their mates. Religious preachers expound on it, and Jesus was perhaps the best known advocate of behavioral change. However, in spite of all the effort from all levels of sources, the nature of behavioral genetics is largely unknown.

Behaviors are the ways in which a person, organism, or group responds to a certain set of conditions. Examples of behavior might be a dog running to fetch a ball, or children pretending to shoot with toy guns. Behaviors also include the responses your body gives to biological changes, such as when you feel stressed. Behaviors can be conscious, unconscious, instinctual, or genetic, or result from habits. Behaviors extend beyond physical display and also includes our emotions and moods. Though an action or emotion is a form of behavior, there is usually no single event that exactly explains a behavior. For example, there is usually no single event that can cause depression or anxiety. Even thinking itself is a behavior, and the kind of thinking that focuses on knowing results in intelligence.

The Human Genome Project

In April 2003 the Human Genome Project, a publicly funded consortium of academic scientists under the direction of Francis Collins, identified the chromosomal locations and structure of the estimated 25,000 genes found within human cells. It was heralded as one of the greatest feats of modern science, and it can be considered akin to some of the greatest adventures of all time—such as putting a man on the moon—but has far greater significance. Some of the genetic applications the HGP made possible are:

- Development of drugs and gene therapy.
- Identification of bacteria and viruses to allow for targeted treatment of disease.
- Identification of which genes control what functions and how they are turned on and off.
- Genetic therapy.
- Understanding of the causes of cancer.
- Genetic engineering.
- Diagnosis and treatment of genetic disorders.

To put this project into perspective, imagine a library of 25,000 books that contain the biological plans for building every living creature, and the knowledge of solutions to diseases that have plagued humans for centuries. However, you can't read the books because they are written in an unknown code of only four letters that are repeated in some mysterious pattern. The secrets of life on earth have been contained in these books since the beginning of life. In this perspective, the books are the genes that have the information to make you, and the library is the human genome. Sequencing genes means learning the order of the four letters that make up the language of

DNA, which is the first step in reading the books of the library.

Begun formally in 1990, the U.S. Human Genome Project was a 13-year effort coordinated by the U.S. Department of Energy and the National Institutes of Health. Project goals were to:

- Identify all the approximately 20,000–25,000 genes in human DNA.
- Determine the sequences of the 3 billion chemical base pairs that make up human DNA.
- Store this information in databases.
- Improve tools for data analysis.
- Transfer related technologies to the private sector.
- Address the ethical, legal, and social issues (ELSI) that may arise from the project.

unraveling the genome

A genome is the entire DNA in an organism, including its genes. Genes carry information for making all the proteins required by all organisms. These proteins determine, among other things, how the organism looks, how well its body metabolizes food or fights infection, and sometimes even how it behaves.

DNA is made up of four similar chemicals (called bases, and abbreviated A, T, C, and G) that are repeated millions or billions of times throughout a genome. The human genome, for example, has 3 billion pairs of bases.

The particular order of As, Ts, Cs, and Gs is extremely important. It underlies all of life's diversity, even dictating whether an organism is human or another species such as yeast, rice, or a fruit fly, all of which have their own genomes and are themselves the focus of genome projects. Because all organisms are related through similarities in DNA sequences, insights gained from nonhuman genomes often lead to new knowledge about human biology.

behavioral genetics

Behavioral genetics is the field of biology that studies the role of genetics in behavior. The field is an overlap of genetics, ethology, and psychology. Classically, behavioral geneticists have studied the heritability of behavioral traits.

Sir Francis Galton (1822–1911) was the first scientist to study heredity and human behavior systematically. The term *genetics* did not even appear until 1909, only two years before Galton's death. With or without a formal name, the study of heredity has always been, at its core, the study of biological variation. Human behavioral genetics, a relatively new field, seeks to understand both the genetic and environmental contributions to individual variations in human behavior. This is not an easy task, for the following reasons.

It is often difficult to *define* the behavior in question. Intelligence is a classic example: Is intelligence the ability to solve a certain type of problem? The ability to make one's way successfully in the world? The ability to score well on an IQ test? During the late summer of 1999, a Princeton molecular biologist published the results of impressive research in which he enhanced the ability of mice to learn by inserting a gene that codes for a protein in brain cells known to be associated with memory. Because the experimental animals performed better than controls on a series of traditional tests of learning, the press dubbed this gene "the smart gene" and "the IQ gene," as if improved memory were the central, or even sole, criterion for defining intelligence. In reality, there is no universal agreement on the definition of intelligence, even among those who study it for a living.

Having established a definition for research purposes, the investigator still must measure the behavior with acceptable degrees of validity and reliability. That is especially difficult for basic personality traits such as shyness or assertiveness, which are the subject of much current research. Sometimes there is an interesting conflation of definition and measurement, as in the case of IQ tests, in which the test score itself has come to define the trait it measures. This is a bit like using batting averages to define hitting prowess in baseball. A high average may indicate ability, but it does not define the essence of the trait.

Behaviors, like all complex traits, involve *multiple genes*, a reality that complicates the search for genetic contributions.

As with much other research in genetics, studies of genes and behavior require *analysis of families and populations* for comparison of those who have the trait in question with those who do not. The result is often a statement of "heritability," a statistical construct that estimates the amount of variation in a population that is attributable to genetic factors. The explanatory power of heritability figures is limited, however, applying only to the population studied and only to the environment in place at the time the study was conducted. If the population or the environment changes, the heritability most likely will change as well. Most importantly, heritability statements provide no basis for predictions about the expression of the trait in question in any given individual.

The following findings indicate that behavior has a biological basis.

- **Behavior is often species-specific.** Some behaviors are so characteristic that biologists use them to help differentiate between closely related species.

- **Behaviors often breed true.** We can reproduce behaviors in successive generations of organisms. (Consider the instinctive retrieval behavior of a yellow Labrador or the herding posture of a border collie.)

- **Behaviors change in response to alterations in biological structures or processes.** For example, a brain injury can turn a polite, mild-mannered person into a foul-mouthed, aggressive boor, and we routinely modify the behavioral manifestations of mental illnesses with drugs that alter brain chemistry.

- **In humans, some behaviors run in families.** For example, there is a clear familial aggregation of mental illness.

- **Behavior has an evolutionary history that persists across related species.** Chimpanzees are our closest relatives, separated from us by a mere 2 percent difference in DNA sequence. We and they share behaviors that are characteristic of highly social primates, including nurturing, cooperation, altruism, and even some facial expressions. Genes are evolutionary glue, binding all of life in a single history that dates back some 3.5 billion years. Conserved behaviors are part of that history, which is written in the language of nature's universal information molecule—DNA.

how genes affect behavior

Traditional research strategies in behavioral genetics include studies of twins and adoptees, techniques designed to sort biological from environmental influences. More recently, investigators have added the search for pieces of DNA associated with particular behaviors, an approach that has been most productive to date in identifying potential locations for genes associated with major mental illnesses such as schizophrenia and bipolar disorder. Yet even here there have been no major breakthroughs, no clearly identified genes that geneticists can tie to disease. The search for genes associated with characteristics such as sexual preference and basic personality traits has been even more frustrating.

In general, it is easier to discern the relationship between biology and behavior for chromosomal and single-gene disorders than for common, complex behaviors that are of considerable interest to specialists and non-specialists alike. So the former are at the more informative end of a sliding scale of certainty with respect to our understanding of human behavior. At the other end of the scale are the hard-to-define personality traits, and somewhere in between are traits such as schizophrenia and bipolar disorder—organic diseases whose biological roots are undeniable, yet unknown, and whose unpredictable onset teaches us about the importance of environmental contributions, even as it reminds us of our ignorance.

Researchers in the field of behavioral genetics have asserted claims for a genetic basis of numerous physical behaviors, including homosexuality, aggression, impulsivity, nurturing, and, recently, spiritualism. A growing scientific and popular focus on genes and behavior has contributed to a resurgence of *behavioral genetic determinism*—the belief that genetics is the major factor in determining behavior.

No single gene determines a particular behavior; behaviors are complex traits involving multiple genes that are affected by a variety of other factors. This fact is often overlooked in media reports hyping scientific breakthroughs on gene function, and, unfortunately, this can be very misleading to the public. For example, a study published in 1999 claimed that over-expression of a particular gene in mice led to enhanced learning capacity. The popular press referred to this gene as "the learning gene" or the "smart gene." What the press didn't mention was that the learning enhancements observed in this study were short-term, lasting only a few hours to a few days in some cases.

Dubbing a gene as a "smart gene" gives the public a false impression of how much scientists really know about the genetics of a complex trait such as intelligence. Once news of the "smart gene" reaches the public, suddenly there is talk about designer babies and the potential of genetically engineering embryos to have intelligence and other desirable traits, when in reality the path from genes to proteins to the development of a particular trait is still a mystery.

With disorders, behaviors, or any physical trait, genes are just a part of the story, because a variety of genetic and environmental factors are involved in the development of any trait. Having a genetic variant doesn't necessarily mean that a particular trait will develop: The presence of certain genetic factors can enhance or repress other genetic factors; genes are turned on and off, and other factors may be keeping a gene from being turned on. In addition, the protein encoded by a gene can be modified in ways that can affect its ability to carry out its normal cellular function.

Human behavioral genetics has been characterized by both excitement and controversy. It has provided more fuel for recent media sensationalism than any other branch of science. Frequent news reports claim that researchers have discovered "the gene" for such traits as aggression, intelligence, criminality, homosexuality, feminine intuition, and even God. Such reports tend to suggest that there is a direct relationship between a gene and behavioral traits or disorders. However, as I've discussed, the reality is that single genes do not mandate most human behaviors.

The understanding that genetic as well as environmental influences affect human behavior is compatible with Darwinian natural selection and, consequently, links human behavior with the evolutionary process. However, behavioral genetics differs from the field of evolutionary psychology in that it focuses on the individual difference affected by genetic influences, rather than the common behavior of a species.

Genomes across species vary very little. For instance, the mouse genome corresponds to the human almost completely, and the gene sequences of the human and the chimpanzee are 99.4 percent alike. The differences are mainly in the activity levels of the genes. For example, certain genes that affect human brain function are much more active in humans than the corresponding genes in the chimpanzee. This disparity is enough to account for the major differences in appearances and behaviors between human and chimp. There is an interesting explanation for why the genomes of different species have so much in common: Scientists propose that all species stem from a single simple organism that existed eons ago. Later species grew off that original species like branches from the trunk of a tree. In every new species, some of the DNA of its predecessors is conserved. This process of change is known as evolution, and the mechanisms by which species emerge include mutation, natural selection, and genetic drift. Mutation is a change in the DNA of a gene that alters the genetic message coded by that gene. Mutation can occur in any part of a gene inside any cell at any point in life. They can be triggered, for example, by radiation, malnutrition, aging, and physical trauma to the cell, although most mutations result from simple errors introduced during replication of the cell.

Behavior results from the genetic coding that occurs in cells through the body, but especially the nervous system, the brain, spine, and a network of nerves through which information is communicated throughout the body, electrically and chemically. Put simply, behavior results from lots and lots of ongoing activity by many, many genes pressed into action by the environment and epigenetic factors. Blood types, some simple metabolic processes, and a few physical traits stem from the action of a single gene, irrespective of environment. Most physical traits and conditions, such as height, blood pressure, weight, and digestive activity, stem from many genes that vary in activity depending on environmental contexts. The same is true for all complex behaviors: Each is affected by multiple genes interacting with multiple environmental influences. Popular opinion is that a gene controls a behavioral trait, period, but the belief that the development of an organism is determined solely by genetic factors is a false belief; scientists have only begun to explore the complex relationship between genetics, environment, and human habits, tendencies, and addictions.

evolutionary psychology

Another method of behavioral genetics study proposes that psychology can be better understood in light of evolution. Specifically, evolutionary psychology proposes that the brain comprises many functional mechanisms called *psychological adaptations* or *evolved psychological mechanisms* (EPMs) that evolved by natural selection. Uncontroversial examples of EPMs include vision, hearing, memory, motor control, and spatial cognition. Most evolutionary psychologists argue that EPMs are universal in a species, excepting those specific to sex or age.

The idea that organisms comprise a number of parts that serve different functions—in other words, living things are, in some sense, machines—goes back at least to Aristotle, and is the foundation of modern medicine and biology. William Paley, drawing upon the work of many others, argued convincingly that organisms are *machines designed to function in particular environments*. Paley believed that this evidence of "design" was evidence for a designer—God. Darwin appears to have been impressed with Paley's argument that organisms are designed for particular environments.

Evolutionary psychology is based on the presumption that, similar to hearts, lungs, livers, kidneys, and immune systems, cognition has a functional structure that has a genetic basis, and therefore has evolved by natural selection. Like other organs and tissues, this functional structure should be universally shared in a species, and should solve important problems of survival and reproduction. Evolutionary psychologists seek to understand cognitive processes by understanding the survival and reproductive functions they might have served throughout the course of evolutionary history. This perspective, though as yet unproven, is reasonable, and compatible with the thought that evolution is God's plan.

sociobiology

In recent decades, another branch of genetics has appeared to study the theory that not only physical traits, but behavior itself, might be inherited. Behavioral geneticists have studied how genes influence behavior, and, more recently, the role of biology in social behavior has been explored. This field of investigation, known as sociobiology, was inaugurated in 1975 with the publication of the book *Sociobiology: The New Synthesis* by American evolutionary biologist Edward O. Wilson. In this book, Wilson proposed that genes influence much of animal and human behavior, and that these characteristics are also subject to natural selection.

Social biologists also study animal behaviors that are called altruistic, meaning unselfish or concerned for others. For example, when birds feed on the ground, one may notice a predator and sound an alarm, and in doing so, draw the predator's attention to itself. Such behavior seems to defy the laws of natural selection, but, in fact, the birds' genes are designed for such behavior. This is also shown by the example of the worker bees' responsibility for collecting food, defending the colony, and caring for the young, while they are sterile and create no offspring. One might wonder, because natural selection rewards those having the highest reproductive success, how a sterile worker could devote itself to others. Well, scientists now recognize that in social insects such as bees, wasps, and ants, the sterile workers are more closely related genetically to one another than in other organisms. Consequently, by helping to protect and nurture their sisters, sterile worker bees preserve their own genes more so than if they actually reproduced themselves. Thus, this altruistic behavior evolved by natural selection. Other examples of animal behavior, including numerous studies on apes and chimpanzees, show similar altruistic behaviors.

Although similar studies on humans are still in process, some suggest that we will find a similar conclusion to the animal studies. Though this has yet to be proven, it is plausible that the altruistic behavior of humans, probably a component of our spiritual genes, is a result of natural selection, and consequently a design for survival of the species.

Sociobiology is often considered a branch of the biology and sociology disciplines, although it uses techniques from a plethora of sciences, including evolution zoology, archeology population genetics, and many others. Within the study of human societies, sociobiology is closely related to the fields of human ecology and evolution psychology. Sociobiology was roundly criticized for contending (without providing much evidence) that genes play a decisive role in human behavior, suggesting there is little we can do to eliminate things such as human aggression. Sociobiologists believe that animal or human behavior cannot be satisfactorily explained entirely by "cultural" or "environmental" factors alone. This belief would then support human spiritual behavior as genetic. They contend that, in order to fully understand behavior, it must be analyzed with some focus on its evolutionary origins. According to Darwin's theory of natural selection, inherited behavioral mechanisms that allowed an organism a greater chance of surviving and/or reproducing would be more likely to survive in present organisms.

It can be argued that the protective behavior of spirituality likely evolved in time because it helped the species survive. Those individuals in the species that did not exhibit such protective behaviors ultimately died out. In this way, spiritual behavior may have evolved in a fashion similar to other types of non-behavioral adaptations.

behavioral genetics and spirituality

As can be assumed by all the varying views, predicting behavioral genetics, or even explaining it, depends on who is speaking. This controversial field, still in its infancy, has a long way to go. Hopefully the work conducted by the Human Genome Project will stimulate sufficient research to provide answers about how we behave, think, and feel, and provide methods of treating disorders and enhancing quality of life.

But for now, all we can do is speculate about the cause and benefits of universal human spirituality. We can reasonably assume that spirituality is a behavior pattern that is a positive for survival, as we have a spiritual nature, and are still here. We can also see that under different environments the altruistic part of spirituality isn't always practiced, such as between competitors for the same resources or position. But when the environment changes and the competition is resolved or compromised, altruism rises to the surface again, at least among common cultures.

In times when surviving the day was the highest level of achievement, ethics were not a high priority. With the development of human consciousness and agriculture's gift of more abundant resources, man was freer to express his innate predisposition for love and compassion to larger communities. From a long-term perspective, it is conceivable that living by humanitarian ethics, instructed by our spiritual genes, may well be the only way humans will eventually survive. But in the meanwhile, the relatively slow pace of evolution will likely be out of sync with the much faster changes in our environment, and our spiritual genetic behavior will not be able to block our prehistoric human instincts and genes that breed violence—that is, until our spiritual nature evolves to a higher genetic priority, or we are able to speed it up with new science, such as epigenetics.

epigenetics

Which is more important in shaping who we are and what we will become—our genes, or the environment around us? For centuries, people have debated whether nature or nurture decides how we look and act. Now, a field of research called epigenetics is showing that we can't really separate one from the other. "We can no longer argue whether genetics or environment has a greater impact on health and development," argues Dr. Randy Jirtle, a genetics researcher in Duke University's department of radiation oncology, "because both are inextricably linked." Jirtle recently co-chaired a conference on epigenetics that was co-sponsored by Duke and NIH's National Institute of Environmental Health Sciences (NIEHS).

Your body "reads" genes and follows their instructions to make you who you are. Epigenetics is the study of factors that can change the way those genes are read without changing the genetic code itself. According to Jirtle, nutrients, toxins, or other things in the environment can cause such "epigenetic" changes. They are one reason, for example, why identical twins who share the same set of genes can be so different.

Epigenetics can have long-lasting effects. Even poor nutrition during pregnancy can cause epigenetic changes that affect a child's risk of getting many diseases later in life. Dr. David Barker of Oregon Health and Sciences University in Portland, Oregon, says, "Studies around the world have shown that people with low birth weight have significantly increased risk of coronary heart disease, stroke, type 2 diabetes, hypertension, and osteoporosis."

Research shows that these epigenetic changes can even be passed down from one generation to the next. Scientists are now discovering exactly how environmental factors produce these lasting changes. A major way genes can be altered is by something called *methylation*. That's when a bunch of atoms scientists call a *methyl group* gets chemically attached to a gene. Dr. Frederick Tyson, a program administrator with NIEHS and co-organizer of the conference, explains, "The addition of the methyl group 'silences' the gene...." In other words, when a gene's had a methyl group attached to it, your body can't read it anymore, even though the genetic code hasn't changed one bit.

NIEHS's research programs now aim to learn more about how environmental factors influence genes and affect your risk for disease. Investigators are looking at a wide range of substances in the environment, including pesticides and heavy metals, to see which can alter genes, and which genes they're altering. At least a dozen drugs targeting epigenetic changes are already being tested, and more are in development. In May 2009, the Food and Drug Administration approved a drug that affects methylation to treat a rare bone marrow disorder that can lead to leukemia. As researchers learn more about how genes and environmental factors interact, they'll be able to develop new approaches for disease prevention and treatment. On a more individual basis, epigenetic mind/body methods can be used to turn off genes that are associated with negative thoughts and instincts. The Natural Happiness 28-Day Program, found in Chapter 9, uses such epigenetic methods.

conscience and our spiritual genes

Conscience is a moral faculty that leads to feelings of remorse when we do things that go against our inbuilt moral precepts. Such feelings are not intellectually arrived at, though they may cause us to examine our conscience and review those moral precepts, or perhaps resolve to avoid repeating the behavior. It is plausible to believe that our conscience is an expression of our spiritual genes—it is a fundamental spiritual genetic command that we follow the good and avoid the bad precepts. This concept is compatible with the genetic behavior mechanism of evolutionary psychology and social biology described previously. What could be more conducive to social behavior than to act in a good (humanitarian) way? However, although the genetic mechanism mandates following good ethics and avoiding bad ones, what's good and what's bad, beyond the moral law, is influenced by our life experiences, environment, culture, and other factors. Conscience can prompt different people in quite different directions, depending on their beliefs, suggesting that, although the capacity for conscience is probably genetically determined, its subject matter is learned, or imprinted, similar to language, as part of a culture. For example, one person can feel a moral duty to go to war; another can feel a moral duty to avoid war under any circumstances.

Conscience, in most religions, is a judgment of reason whereby the human person recognizes the moral quality of a concrete act he or she is going to perform, is in the process of performing, or has already completed. Obedience to conscience has been claimed by many dissenters as a God-given right; for example, from Martin Luther, who said (or reputedly said), "Here I stand, I can do no other," in disagreement with certain doctrines and Catholic dogma. The Church eventually agreed, saying, "Man has the right to act according to his conscience and in freedom so as personally to make moral decisions. He must not be forced to act contrary to his conscience, nor must he be prevented from acting according to his conscience, especially in religious matters." Many religions consider following one's conscience to be as important as, or even more important than, obeying human authority. This can sometimes lead to moral quandaries: "Do I obey my spiritual nature, church/military/political leader, or do I follow my own sense of right and wrong?"

Because our spiritual nature is genetic, as is our belief in God, it would seem to be the highest ethics over human authority, but then, to each their own.

genetic priorities

Our behavior is strongly influenced by the priority of our genes and instincts, which is the inherent disposition of a living organism toward a particular behavior. Instincts are generally inherited patterns of responses or reactions to certain kinds of stimuli, or biological necessities. In humans they are most easily observed in behaviors such as emotions, sexual drive, and other bodily functions, as these are largely biologically determined. Instinct provides a response to external stimuli, which moves an organism to action unless overridden by intelligence or beliefs. Our instincts have a priority, depending on biological necessities such as survival, security, and procreation.

For instance, changes in environment, along with perception, can determine whether survival is an issue or not. If a change in environment generates a perception of a threat to a biological requirement, such as survival, this perception would normally have a first priority on our behavior. Other priorities would be determined by biological necessity, or, in some cases, beliefs. The priority of behavior is then a sequence requiring the highest priority to be satisfied before the next priority becomes the highest, and so on. For example, the priority for physical survival (food, water, sleep, and so on) must be satisfied before the next priority for safety (protection) becomes the motivator. These instinctual priorities explain much about human behavior throughout the ages. For example, our prehistoric ancestors lived in an environment where survival was a full-time effort, and all other genetic drives were sublimated. Later, when agriculture came into being, and the food supply became more predictable, the next priority, security, became the issue. As man formed into communities, protection became less of a concern, and man turned his attention to self-transformation. Spirituality was expressed in a belief in a God who commanded the ethics of compassion, honesty, and justice, and a conscience that punished for "bad" actions.

It is debatable whether or not living beings are bound absolutely by instinct. Though instinct is what seems to come naturally or perhaps with genetic heredity, general conditioning and the environment surrounding a living being play a major role. Predominately, instinct is pre-intellectual, whereas intuition is trans-intellectual.

Abraham Maslow developed the Hierarchy of Needs model in 1940s–50s United States, and his theory remains valid today for understanding human motivation. Maslow developed his model based on well-known individuals and statistical studies; his hierarchy is another way of describing human instincts. However, even while our highest priority instinct for survival is being threatened, our spirituality is expressed in our plea for help from God, when we perceive that we are in serious danger. A well-known phrase for this interaction, even from supposedly non-spiritual people, is that "there are no atheists in foxholes."

ranking of genetic priorities

Generally each priority must be satisfied before the next lower one is expressed. Each condition depends on subjective perception; strong beliefs can override any instinct—for instance, in the present environment, suicide bombers.

#1: Survival. The priority of air, food, sleep, sex, and other biological requirements to survive and propagate the species.

#2: Security. When the survival priority is met, the security priority comes into play, finding safe circumstances, stability, power, and protection.

#3: Belonging. After the priorities of survival and security are, by and large, taken care of, the priority of belonging is enabled. This relates to individuals and groups, such as friends, a sweetheart, spouses, children, affectionate relationships in general, a community, and a church.

#4: Self-fulfillment. After the base biological needs are met, the genetics for self-fulfillment are enabled. Self means basically becoming everything you have the potential and desire to be. Material self-fulfillment would include such feelings as confidence, competence, achievement, power, fame, independence, and freedom. As the individual matures in life experiences, spiritual self-fulfillment of love, compassion, service, and hope become more emotionally rewarding. The zeal for spirituality can also be affected by certain genetic mutations.

#5: Self-transcendence. After spiritual self-fulfillment has progressed sufficiently, the path to self-transcendence is cleared, in which one transcends the self to a feeling of oneness with the world. Universal compassion becomes natural, and the peak experiences of communion with the creator bring moments of ecstasy and joy. It may seem strange that humans are genetically programmed to experience all of the previous ego-based priorities of individual survival and material self-fulfillment, only to reverse our direction and prepare for transcending our individuality to be at one with the world. Although this behavioral change may not seem logical, we know that this is the human design, by the proof of the rewarding emotions of spiritual expression.

Characteristics of self-transcendence include:

- Unconditional compassion.
- Unconditional love.
- Inner peace, unthreatened and unworried by the known and unknown.
- A sense of oneness.
- Detachment from materialism and worldly desires.
- Living in the moment.
- Peak experiences—feelings of ecstasy, love, and joy.
- Unity of the universe and all it contains.

spirituality and evil

Every day news reports tell of horrific, sad, or deeply disturbing events worldwide, and we wonder why such things happen. Is it humanity's own doing? Is it evil at work?

Before we can understand why evil exists, we need to understand what evil *is*. Most dictionaries give synonyms such as *morally wrong*, *causing great harm*, *wickedness*, *sin*, and *depravity*. Here are some definitions of evil frequently given:

1. A mistake made by basically good and decent people.
2. A natural response of our deeply sinful basic natures.
3. The result of the devil's seductions for basically weak people.
4. An illusion.
5. A mistake in our past life.
6. Nonexistent.
7. Proof that there is no God.

For Judaism, the belief is #1: Evil is basically a mistake made by people who have been given the freedom to choose between good and evil, life and death, a blessing and a curse, and who sometimes tragically choose poorly.

For Christianity, Islam, and Zoroastrianism, the main beliefs are #1, #2, and #3: People's sinful nature is the reason why Christ died, which atones for humankind's sins and is so crucial to salvation for Christians. For Muslims, the beliefs are mainly #1 and #3: In addition to being good, people are also prideful and easily seduced by the devil into forgetting their rightful place. For Hinduism and Buddhism, the beliefs are #4, #5, and #6: Evil is just one of many illusions we harbor because of our low level of spiritual understanding. Atheists often choose #7; indeed, for many people, the presence of evil in the world is the main reason they don't believe in God or religion.

Polytheistic religions, such as Hinduism, explain that evil comes from the bad gods and goddesses, and goodness comes from the good gods and goddesses. In monotheistic religions, evil has to be explained in such a way that its existence doesn't negate people's beliefs about an omnipotent God who is infinitely good. Because, if God can rid the world of evil but won't, He *isn't* infinitely good. Moreover, if He *would* rid the world of evil but can't, He isn't omnipotent. A common response to this is that evil doesn't come from God but from the choices and actions of the beings He created.

Before modern genetic science, insufficient knowledge was available to provide a reasonably objective reason for evil. Consequently, past explanations for evil are subjectively biased by the source—usually religious—supplying the opinion.

environmental modifications
of spiritual programming

From a genetic priority view, evil behavior and its cause has much to do with genetic priorities and the existing personal environment. Under certain circumstances—for example, with your back against the wall and threatened—a violent response even from a highly ethical person is a biological and psychological instinct. Killing people, except in self-defense, is condemned by virtually all religions and the great majority of persons. However, in times of war, regardless of the reason, killing is not only condoned but in many cases is celebrated for "the good of the cause." Some churches' teachings incite man to confess his evil nature, but the fact is that man is *not* responsible for his nature; it is part of his original equipment, integral to his DNA.

In ancient times, the struggle for survival took priority over humanitarian ethics. In fact, ethics were not a consideration, or even known. Even in more "recent" writings, such as the Old Testament, man consistently rejected the ways of God for his own selfishness. In modern times, with survival and security less of a issue, as it has been jokingly said, man can afford a better set of ethics. In our modern environment, a combination of more resources and education has provided spirituality with more freedom to express itself.

Even as spirituality evolves and is being expressed more, the environment is changing even faster. Because behavior is a result of genetic/environmental interaction, when the environment becomes hostile, violence results. For example, when one entity believes that its security is being threatened, violence results in conflicts, perhaps even escalating into wars. In other words, man reverts to his high-priority instincts as he perceives his changing environment.

In today's world, violence and injustice appear to stem from the fear of the security of individuals, groups, and countries being jeopardized. For example, when the interests of one country seem to threaten the security of another, higher genetic priorities are enacted. The only realistic way to keep peace is to increase the spiritual expression of the ethics of oneness and compassion. Such a scenario seems a long way off, but hints that it can happen are shown in the random acts of compassion that surface in hostile environments.

The position of this book is that the evolutionary design of our Creator includes the following design concepts:

- Humans are not born with a sinful nature (that would be inconsistent with a beneficent Creator). Rather, the evolutionary design of our Creator equips humans to adapt to a changing environment. Because the environment has changed faster than evolution can adapt, certain instincts become out of sync.

- Humans are all born with virtually identical genetics for survival, safety, love and connectedness, self-fulfillment, and self-transcendence. The priority of expression of these instincts is a function of our environment and beliefs.

- Evil is behavior that another perceives injures his or her rights.

- Evil is only evil when the perpetrator knows his or her action is bad.

- Conflicting interpretations of God's will can result in evil and violence.

- Conflicting religious beliefs have long been a source of evil.

- Evil frequently results from genetic priority being revised by a changed environment.

- The behavior of all normal humans is not premeditated, but responsive to their genetic programming and their environment.

- Humanitarian ethics, such as love, compassion, and service, are an expression of our spiritual genes.
- The ultimate goal of all human behavior is rewarding emotions and pain avoidance.
- Our Creator designed our genetic behavior and motivated us with good emotions.
- Unique behavior is always possible due to genetic mutations and malfunctions.

conclusion

If it is the case that we are endowed with spiritual genes that prompt a belief in a supernatural power and spiritual ethics, what is the answer to why evil and violence continue to exist in the world? This age-old question has been worked throughout the ages by a long list of theologians, thinkers, activists, and authorities of every type, without reaching a time-proven answer.

Could it be that that the reason as to why there is evil in a spiritual world is the result of the priority of human instincts?

appendix II

Self-Hypnotic Induction

Record the following script and play it back to yourself in order to relax into a hypnotic state. For self-hypnosis, assume the guided instructions are yourself speaking to you. (Do not read aloud the words in brackets.)

Get yourself in a comfortable sitting position, somewhere you won't be disturbed.

Close your eyes and perform the breathing exercise you practiced for mindfulness in Chapter 10. Focus on your breathing. Feel your breath coming in your left nostril and then out your right nostril. Breathing normally, be conscious of your breathing.

- Breathe in to the count of 1 and out to the count of 2. [2 seconds.]
- In to the count of 3 and out to the count of 4. [2 seconds.]
- In to the count of 5 and out to the count of 6. [2 seconds.]

- In to the count of 7 and out to the count of 8. [2 seconds.]
- In to the count of 9 and out to the count of 10. [2 seconds.]

We will stop while you repeat this breathing exercise yourself. [Stop for 15 seconds.]

Now stop your counting and concentrate on my voice, while you continue your rhythmic breathing.

Invite your body to release any tension, and imagine the tension is melting away as you focus on each part of the body.

- Invite your feet and ankles to relax and notice how they respond. [Wait 2 seconds.]
- Imagine your legs are completely relaxed. [Wait 2 seconds.]
- Now feel the tension melt out of your hips. [Wait 2 seconds.]
- Feel the tension dissolve in your body as you feel your abdomen relax.
- Now feel the softness flow up from your abdomen up to your chest. Relax. [Wait 2 seconds.]
- Now feel your arms relax and become soft. [Wait 2 seconds.]
- Finally, imagine all the tension dissolves from your neck, so that your whole body is relaxed, and all the stress and tension has dissolved. You are now in a deep state of relaxation. [Wait 3 seconds.]

In a moment you will count down from 5 and you will imagine your favorite place. It can be anywhere that is very beautiful and private, where you feel relaxed and peaceful. [Wait 3 seconds.]

It could be in a beautiful garden, where you can smell the wonderful aromas wafting on the light breeze, while you feel warmed by the sun and alone with the beauty of nature. [Wait 3 seconds.]

Or it could be on a sandy beach with the blue waves breaking on the shore and the sandpipers skirting the changing water edge. And you can smell the aromas and hear the sounds of the surf and feel the sand between your toes. [Wait 2 seconds.]

Now we start to count down to get to your favorite place.

- 5—And you relax deeper.
- 4—You are feeling more peaceful.
- 3—Feel how comfortable you are; soon you will be in your favorite place.
- 2—Your breathing is regular.
- 1—You are almost there.
- 0—It is so good to be here. [Wait 2 seconds.]

You are so glad to be in your favorite place where you are so relaxed and have not a care in the world. Take some time to use all your senses to fully imagine your favorite place. You may hear the sounds and smell the aromas or even imagine a favorite taste. [Wait 5 seconds.]

Listen to the voice and believe the suggestions we are going to give you will make you happier and healthier and you will enjoy life so much more. You will follow them without effort.

- Imagine the disadvantages of your present materialistic behavioral habits. [Wait 2 seconds.]
- See how they are making you stressed and threatening your health. [Wait 2 seconds.]
- Imagine how your selfish attitude makes you appear to others. [Wait 2 seconds.]
- Imagine all the energy it takes to get through the day with all those negative thoughts. [Wait 2 seconds.]
- You feel powerless to control all those random thoughts that build up stress.
- Now imagine how having a peaceful mind will make you feel. [Wait 2 seconds.]
- Imagine that you are feeling healthier and more energetic. [Wait 2 seconds.]
- Imagine being able enjoy those circumstances you couldn't stand before [Wait 2 seconds.]
- Imagine how your relationships will get better. [Wait 2 seconds.]
- Imagine how wonderful it will feel to be compassionate with others [Wait 2 seconds.]

You want to get rid of the unhealthy habits of negative thoughts and competition for resources that make you worry and build up stress that negatively affects your health, but you are unable to break these habits. You think you are unable to break your unpleasant habits because they are rooted in your genes and childhood environment. But now you believe that by expressing spiritual acts you can become a loving, compassionate person.

Now imagine your present self as a **dark outline** of an unloving figure projected on a screen.

Now imagine your new self as a loving, thinner white outline overlaying the dark outline.

Imagine the new white outline is becoming larger... and larger...until you cover more...and more of the unloving dark outline you used to be. Whenever you think of the words **LOVING FIGURE** you will see this image in your mind's eye.

Now repeat the following affirmations after me. [Pause for 2 seconds after each.]

- I can change my selfish behavior to be unconditionally loving.
- I can satisfy my need for others by compassionate acts.
- I will practice spiritual acts daily, until they become habits.
- I will make the peace of mindfulness the foundation of my life.
- Self-seeking pleasure is a poor substitute for natural happiness.
- Natural happiness always works; pleasures don't.
- I am not worried about what others think about my spiritual attitude.
- I know that more acts of relaxation response will make me healthier and more energetic.
- I enjoy materialistic pleasures, but I know that overuse leads to disappointment.
- I am enjoying life more and more as I become more compassionate.
- I know that the will of God is agape love.

- I know materialism is not a substitute for love or natural happiness.
- I know that caring for others brings more joy than the amount of my stuff.
- As I become more loving, I enjoy life more and more and feel alert and alive.

Now is the time to implant happiness habit suggestions in your subconscious, which will remain with you after you have awakened.

Repeat after me:

- If I do selfish acts, I will put my hand on my head and think **LOVING FIGURE**, imagining the thinner outline of my white figure against my old, larger, dark figure.
- If I want materialistic pleasures at an inappropriate time I will place my hand on my head and think the words **LOVING FIGURE** and imagine my white figure outline.

Now let the image of your favorite place fade, as you come up to the surface of your mind as I count up to five. [Pause for 2 seconds.]

You will feel calm and relaxed as you know you can return whenever you want.

- **Zero**—Become aware of your breathing.
- **One**—If you want unhealthy materialistic pleasure, you will place your hand over your head and think, **LOVING FIGURE.**
- **Two**—You are coming back to the surface and will become mindful.

- **Three**—Whenever you are thinking about materialistic acts, you will imagine your white **LOVING FIGURE** overshadowed by your larger, former dark figure outline.

- **Four**—Whenever you think about your favorite place you will think about LOVING ACTS.

- **Five**—Take a deep breath, inhaling deeply and exhaling, as you have returned to the surface of your mind. You are fully awake and refreshed. Take a moment to adjust to your surroundings with a brief mindful breathing exercise counting up to 10. [Wait 4 seconds.]

Remember this suggestion: "Each time I listen to this exercise, I can relax even more deeply."

✳ ✳ ✳

Now there is nothing stopping you from relaxing even more comfortably while you discover new strategies for natural happiness!

Notes

Chapter 1

1. Nobel Prize–winners from Cold Spring Harbor Science Walking Tour (Cold Spring Harbor Department of Public Affairs Publication (undated, page 2). In 1969, Alfred Hershey, Slavador Luria, and Max Delbruck won the Nobel Prize for proving that DNA is inherited, and for their research in bacterial genetics. Their research showed that bacteria can become resistant to antibiotic medicine. Barbara McClintock won her Nobel Prize in 1983 for jumping genes called *transposons*, which she observed while studying the DNA of maize. In 1993, Richard Roberts and Philip Sharp demonstrated RNA splicing and how genes can split, earning them both the coveted Nobel Prize in Science. In 2009, Carol Greider and her colleagues Elizabeth Blackburn and Jack Szostak won the Nobel Prize for Physiology or Medicine.

2. "Eugenics."

3. DNA Initiative.

4. Ornish, et al. "Changes in prostate gene."

5. "What Is Epigenetics?"

6. Ibid.

7. "What Is Epigenetics?"

8. "Epigenetics means."

9. Ibid.

10. Ibid.

11. Alleyne, "Childhood Abuse."

12. Ibid.

13. "About the Human Genome Project."

14. Ibid.

15. Ibid.

16. Dawson Church, interview with Laurie Nadel, July 30, 2009.

Chapter 2

1. Crews, definition.

2. "Who's Happy and Why?"

3. Ibid.

4. Nixon, "Epigenetics."

5. Ibid.

6. Nadel, Haims, and Stempson, *Dr. Laurie Nadel's*, p. 210.

7. Ibid.

8. Mitchell, interview.

9. Braden, interview.

10. Braden, *Spontaneous Healing*, and interview.
11. Ibid.
12. Ibid.
13. Livescience staff, "Happiness."
14. Ibid.
15. McCraty, interview.
16. Ibid.
17. Ibid.
18. Ibid.
19. Ibid.
20. Ibid. (250mm is the specific wavelength that DNA absorbs.)
21. Thanks to Cynthia Larson (*www.realityshifters.com*) for her signature question!

Chapter 4

1. Doyle, "Botswana."
2. Crockett, *Stone Age*, p. 95.
3. "Spirituality," Dictionary.com.
4. Guralnik, *Webster's*, p. 576.

Chapter 5

1. Hagerty, *Fingerprints*, p. 97–8.
2. Pert, *Molecules*, p. 131–32.
3. Ibid., p. 98.
4. Ibid., p. 98.
5. Ibid., p. 101.

6. Braden, interview.

7. Ibid.

8. Schwarz and Simon, *G.O.D. Experiments*, p. 65–6.

9. Braden, interview.

Chapter 6

1. Informal survey conducted by Dr. Laurie Nadel, 1997–2000, among corporate executives attending her critical-thinking seminars.

2. Niemiec, Ryan, and Deci, "The path taken."

3. Grove and Prince, *Facts & Figures*.

4. Buckingham, "What's happening."

5. Ibid.

6. Veenhoven, "Happiness in Nations," and "World Database."

7. Veenhoven, "World Database."

8. "The World Map of Happiness."

9. "Map of World Happiness."

10. "The World Map of Happiness."

Chapter 7

1. "Happiness."

2. "Who's Happy and Why."

3. Ibid.

4. "Mood May Be Written."

5. Jablonka and Lamb, *Evolution*, p. 1.

6. "Symptoms," p. 168.

7. "National Accounts."

8. Moreira-Almeida, "Religiousness."

9. Quote from *www.thinkexist.com*.

10. Czikszentmihalyi, "Finding Flow."

11. Pattakos, *Prisoners*.

12. Puchalski, "Spirituality."

13. Harris, Lufkin, Benisovich, Standard, Bruning, Evans, and Thoresen, "Effects of a group forgiveness."

14. Ibid.

15. Kent, "The Cure for Depression?"

16. Gardner, *A Thousand Clowns*, p. 75.

Chapter 8

1. Collins, *The Language of God*.

2. Baird, *The Mindful Meals*.

Chapter 9

1. Nadel, "Make Way."

2. Kent, "The Cure for Depression?"

3. Braden, interview.

4. Kornfield, *The Roots*.

Glossary

agape love: One of the words for the kinds of love in the Greek language of the New Testament. Agape love is of the will, meaning to seek the highest good for your neighbor. *Agape* was the translation of *the love* that Jesus used. Because agape love is of the will, and not necessarily emotional, it makes Jesus' command to "love your enemy" realistic.

altruism: Unselfish concern for the welfare of others. It is a traditional virtue in many cultures, and a core aspect of various religious traditions such as Judaism, Christianity, Islam, Hinduism, Jainism, Buddhism, Confucianism, and Sikhism. This idea was often described as the "Golden rule of ethics." Altruism is the opposite of selfishness, and focuses on a motivation to help others or a desire to do good without reward.

behavior: *Behavior* refers to the actions or reactions of an object or organism, usually in relation to the environment. Behavior can be conscious or subconscious, overt or covert, and voluntary or involuntary. Human behavior can be common, unusual, acceptable, or unacceptable. Humans evaluate the acceptability of behavior by using social norms, and regulate behavior by means of social control.

behavior genetics: The study of the interaction of heredity and environment insofar as they affect behavior. The question of the determinants of behavior is commonly called the "nature-nurture" controversy. A balanced view that recognized the importance of both genetics and environment prevailed in the 1970s, and modern research is focused on identifying genes that affect behavioral dimensions, such as personality, intelligence, and disorders like depression and hyperactivity.

It is important to note that there is no single gene for intelligence, personality traits, spirituality, behavior, or even height. Rather, such complex characteristics are *polygenic*—they are influenced by multiple genes. The research methodologies do not tell us *which* genes are involved, only the relative influence of all genes as opposed to environment.

biological altruism: In the field of evolutionary biology, an organism behaves altruistically when its behavior benefits other organisms, at a cost to itself. The costs and benefits are measured in terms of *reproductive fitness*, or expected number of offspring. So, by behaving altruistically, an organism reduces the number of offspring it is likely to produce itself, but boosts the number that other organisms are likely to produce.

The biological notion of altruism is different from the common concept of altruism, which involves the conscious intention of helping another. In the biological sense there is no such requirement, because some examples of biological altruism are found among creatures that are (presumably) not capable of conscious thought at all, such as insects. For the biologist, it is the consequences of an action for reproductive fitness that determine whether the action counts as altruistic, not the intentions, if any, with which the action is performed.

Altruistic behavior is common throughout the animal kingdom, particularly in species with complex social structures. For example, vampire bats regularly regurgitate blood and donate it to other members of their group who have failed to feed that night, ensuring that they do not starve. In numerous bird species, a breeding pair receives help in raising its young from other "helper" birds, which protect the nest from predators and help to feed the fledglings. Vervet monkeys give alarm calls to warn fellow monkeys of the presence of predators, even though in doing so they attract attention to themselves, increasing their personal chance of being attacked.

biological imperatives: The needs of living organisms required to perpetuate their existence; to survive. They include the following hierarchy of logical imperatives for a living organism: survival, territorialism, competition, reproduction, and quality of life–seeking.

Competition is one of the more significant environmental factors that constitute natural selection. Individual organisms compete for food and mates; groups of living organisms compete for control of territory and resources.

compassion: A human emotion elicited by the suffering of others. The feeling commonly gives rise to an active desire to reduce another's suffering. It is a key component in the spiritual expression of altruism. It has been manifested down through the ages by the so-called Golden Rule: *Do to others what you would have them do to you.*

competition: A contest among individuals, groups, nations, animals, and such for territory, a niche, power, mates, or limited resources. It arises whenever two or more parties strive for a goal that cannot be shared. Competition occurs naturally between living organisms that coexist in the same environment. For example, our ancestors instinctively competed for water, food, and mates, and in later times, when these needs were met, deep rivalries often arose over the pursuit of wealth, prestige, and fame.

Throughout history, competition has fueled wars and destruction, but it also may give incentives for self-improvement. For example, if two companies are competing for the same market, they will improve their products and service to increase sales. If one company is more responsive to the needs of consumers, this company will flourish. In addition, the level of competition can also vary; at some levels, competition can be informal—more for pride and/or fun. However, other competitions can be extremely serious; for example, some human wars have erupted because of intense competition between two nations. Competition can have both beneficial and detrimental effects. Many evolutionary biologists view inter-species and intra-species competition as the driving force of adaptation, and, ultimately, of evolution.

consciousness: Generally defined as a subjective experience or awareness or wakefulness, or the executive control system of the mind. Beyond that, consciousness as yet refuses to be defined, and it is an umbrella term that may refer to a variety of mental phenomena. For the purposes of this book, consciousness, unlike unconsciousness, is that part of the mind that permits spiritual expression.

Although science has agreed on some fundamental theories of how consciousness works, it is a long way from actually determining what consciousness *is*.

Because our brain is a "work in progress" of evolution, it is still evolving, and as such has a continuum of limitations. At any rate, mainstream neuroscientists are convinced that consciousness is the product of brain activity. However, at the present, we can't comprehend how the data processing of our brain cells relates to our subjective experiences.

chromosome: In the nucleus of each cell, the DNA molecule is packaged into thread-like structures called *chromosomes*. Each is made up of DNA tightly coiled many times around proteins called *histones* that support its structure. Chromosomes are not visible in the cell's nucleus—not even under a microscope—when the cell is not dividing. However, the DNA that makes up chromosomes becomes more tightly packed during cell division and is then visible under a microscope. Most of what researchers know about chromosomes was learned by observing chromosomes during cell division.

Each chromosome has a constriction point called the *centromere*, which divides the chromosome into two sections, or "arms." The short arm of the chromosome is labeled the *p arm*. The long arm of the chromosome is labeled the *q arm*. The location of the centromere on each chromosome gives the chromosome its characteristic shape, and can be used to help describe the location of specific genes.

DNA: a.k.a deoxyribonucleic acid; the hereditary material in humans and almost all other organisms. Nearly every cell in a person's body has the same DNA. The information in DNA is stored as a code made up of four chemical bases: adenine (A), guanine (G), cytosine (C), and thymine (T). Human DNA consists of about 3 billion bases codes, and more than 99 percent of those are the same in all people. The order, or sequence, of these bases determines the information available for building and maintaining an organism, similar to the way in which letters of the alphabet appear in a certain order to form words and sentences.

DNA bases pair up with each other, A with T and C with G, to form units called base pairs. Each base is also attached to a sugar molecule and a phosphate molecule. Together, a base, sugar, and phosphate are called a *nucleotide*. Nucleotides are arranged in two long strands that form a spiral called a *double helix*. The structure of the double helix is somewhat like a ladder, with the base pairs forming the ladder's rungs and the sugar and phosphate molecules forming the vertical sidepieces of the ladder.

An important property of DNA is that it can replicate, or make copies of itself. Each strand of DNA in the double helix can serve as a pattern for duplicating the sequence of bases. This is critical when cells divide, because each new cell needs to have an exact copy of the DNA present in the old cell.

emotion: No aspect of our mental life is more important to the quality and meaning of our existence than emotions; they are what make life worth living. Generally an emotion is defined as a mental and physiological state associated with a wide variety of feelings, thoughts, and behavior. Emotions are subjective experiences, often associated with mood, temperament, personality, and disposition. In this book we are speaking about the rewarding emotions, such as joy and love, that stem from spiritual expressions such as compassion.

empathy: The capability of sharing another being's emotions and feelings. Because empathy involves understanding the emotional states of other people, the way it is characterized is a derivative of the way emotions themselves are characterized. If, for example, emotions are taken to be centrally characterized by bodily feelings, then grasping the bodily feelings of another will be central to empathy. On the other hand, if emotions are more centrally characterized by a combination of beliefs and desires, then grasping these beliefs and desires will be more essential to empathy.

The ability to imagine oneself as another person is a sophisticated imaginative process. However, the basic capacity to recognize emotions is probably innate and may be achieved unconsciously. The human capacity to recognize the bodily feelings of another is related to one's imitative capacities, and seems to be grounded in the innate capacity to associate the bodily movements and facial expressions one sees in another with the feelings of producing those corresponding movements or expressions oneself.

Research in recent years has focused on the possible brain processes underlying the experience of empathy. For instance, functional magnetic resonance imaging (fMRI) has been employed to investigate the functional anatomy of empathy. These studies have shown that observing another person's emotional state activates parts of the neuronal network involved in processing that same state in oneself, whether it is, touch, pain, or disgust.

energy: In physics, energy is a scalar physical quantity that describes the amount of work that can be performed by a force—an attribute of objects and systems that is subject to a conservation law—which means that the total energy always remains the same. Different forms of energy include kinetic, potential, thermal, gravitaional, sound, light, elastic, and electromagnetic. The forms of energy are often named after a related force, and any form of energy can be transformed into another form or system. In biology, the energy of our thoughts is reflected in our emotions, and in mind/body innovations the energy of our mind is transformed into biological changes.

epigenetics: Refers to changes in phenotype (appearance) or gene expression caused by mechanisms other than changes in the underlying DNA sequence. These changes may remain through cell divisions for the remainder of the cell's life and may also last for multiple generations. However, there is no change in the underlying DNA sequence of the organism; instead, non-genetic factors cause the organism's genes to behave (or express themselves) differently. Epigenetics is the study of these reactions and the factors that influence them.

The genome dynamically responds to the environment. Stress, diet, behavior, toxins, mind/body interventions, and other factors activate chemical switches that regulate gene expression. The mind-body interventions used in this book are forms of epigenetics mechanisms.

evolution: Change in the genetic material of a population of organisms throughout a long period of time. Although the changes in a generation are small, throughout a long period of time they accumulate, to cause significant changes that can result in new species. Commonalities among species indicate that all known species descended from a common ancestor.

The basic process of evolution is the passing of genes from one generation to the next, resulting in an organism's inherited traits. These traits vary within populations. Evolution is the result of processes that constantly introduce variation and processes that make those variants become either more common or more rare. New variation arises in two main ways: either from mutations in genes or from the transfer of genes between populations and between species. *Natural selection* is the process of choosing traits that increase the chance of survival and reproduction, and causing them to become more common in a population. Another mechanism of evolution is *genetic drift*, a process that produces random changes, resulting in chance having an effect on whether a given trait will be passed on to future generations.

evolutionary psychology: Attempts to explain psychological traits, such as memory, perception, consciousness, or language as adaptations. *Adaptations* are functional products of natural selection that helped our ancestors get around the world, survive, and reproduce.

Other adaptations might include the ability to infer others' emotions, to discern kin from non-kin, to identify and prefer healthier mates, and to cooperate with others. Consistent with the theory of natural selection, evolutionary psychology sees organisms as often in conflict with others of their species, including mates and relatives. Humans, however, have a marked capacity for cooperation under certain conditions as well.

faith: The confident belief in a person, idea, or thing. In a religious context, the word *faith* frequently refers to God or a religion itself. Faith involves a concept of future events or outcomes, and is used conversely for a belief not resting on logical proof or material evidence. Informal usage of the word *faith* can be quite broad, and may be used in place of *belief*.

There is a difference of opinion with respect to the initial meaning of *faith*. St. Augustine of Hippo holds that all of our beliefs rest ultimately on beliefs accepted by faith. Others, such as C.S. Lewis, hold that faith is merely the virtue by which we hold to our reasoned ideas, despite moods to the contrary.

feeling: In psychology, the word *feeling* is also used to describe experiences other than the physical sensation of touch, such as a feeling of warmth. In psychology the word is usually reserved for the conscious subjective experience of emotion. Many schools of psychotherapy depend on the therapist achieving some kind of understanding of the client's feelings, for which different methodologies exist. Some theories of interpersonal relationships also have a role for shared feelings or the understanding of another person's feelings.

Perception of the physical world does not necessarily result in a universal reaction among receivers, but instead varies depending on one's tendency to handle the situation, how the situation relates to the receiver's past experiences, and any number of other factors. Feelings are also known as a state of consciousness, such as that resulting from emotions, sentiments, or desires.

gene: The basic physical and functional unit of heredity. Genes, which are made up of DNA, act as instructions to make molecules called *proteins*. In humans, genes vary in size from a few hundred DNA bases to more than 2 million bases. The Human Genome Project has estimated that humans have between 20,000 and 25,000 genes.

Every person has two copies of each gene, one inherited from each parent. Most genes are the same in all people, but a small number of genes (less than 1 percent of the total) are slightly different between people. *Alleles* are forms of the same gene with small differences in their sequence of DNA bases. These small differences contribute to each person's unique physical features.

gene expression: The process by which information from a gene is used in the synthesis of a functional gene product. These products are often proteins, but in non-protein coding genes, the product is a functional RNA to generate the macromolecular machinery for life.

244   H<small>APPINESS</small> G<small>ENES</small>

In genetics, gene expression is the most fundamental level at which genotype gives rise to the phenotype. The genetic code is "interpreted" by gene expression, and the properties of the expression products give rise to the organism's phenotype.

gene mutation: A permanent change in the DNA sequence that makes up a gene. Mutations range in size from a single DNA building block (DNA base) to a large segment of a chromosome. Gene mutations occur in two ways: They can be inherited from a parent or acquired during a person's lifetime. Mutations that are passed from parent to child are called *hereditary mutations* or *germline mutations* (because they are present in the egg and sperm cells, which are also called germ cells). This type of mutation is present throughout a person's life in virtually every cell in the body.

Mutations that occur only in an egg or sperm cell, or those that occur just after fertilization, are called *new* (*de novo*) *mutations*. De novo mutations may explain genetic disorders in which an affected child has a mutation in every cell, but has no family history of the disorder.

Acquired (*somatic*) *mutations* occur in the DNA of individual cells at some time during a person's life. These changes can be caused by environmental factors, such as ultraviolet radiation from the sun, or can occur if a mistake is made as DNA copies itself during cell division. Acquired mutations in somatic cells (cells other than sperm and egg cells) cannot be passed on to the next generation.

genetics: The study of genes, their function, structure, and effect. Genes are the means by which living organisms inherit features from their ancestors; for example, children often look similar to their parents. In genetics, a feature of an organism is called a *trait*. Some traits are features of an organism's physical appearance, but there are many other trait types, and these range from aspects of behavior to resistance to disease. Traits are often inherited—for example, tall and thin people tend to have tall and thin children. Other traits come from the interaction between inherited features and the environment. For example, a child might inherit the tendency to be tall, but if little food is available and the child is poorly nourished, it will still be short. The way genetics and environment interact to produce a trait can be complicated; for example, the chances of somebody dying of cancer or heart disease seem to depend on both his or her family history and lifestyle.

Genetic information is carried by a long molecule called DNA, which is copied and inherited across generations. Traits are carried in DNA as instructions for constructing and operating an organism. These instructions are contained in segments of DNA called *genes*. DNA is made of a sequence of simple units, with the order of these units spelling out instructions in the genetic code. This is similar to the order of letters spelling out words. The organism "reads" the sequence of these units and decodes the instruction.

Not all the genes for a particular instruction are exactly the same. Different forms of one type of gene are called different *alleles* of that gene. As an example, one allele of a gene for hair color could carry the instruction to produce a lot of the pigment in black hair, while a different allele could give a garbled version of this instruction, so that no pigment is produced and the hair is white.

genetic engineering: Used to take genes and segments of DNA from one species, such as a mouse, and transpose them into another species, such as a monkey or a banana. In very simplistic terms, this is performed by cutting loose strips of DNA and placing them randomly or in a number of specific sites.

Genetic engineering makes it possible to break through the species barrier and to shuffle, for example, an insect-killing toxin gene from bacteria into maize, cotton, or rape seed—or, for that matter, genes from humans into a pig.

As straightforward as it sounds, the theory doesn't hold up with reality. Often, for no apparent reason, the new gene only works for a limited amount of time and then "falls silent." But there is no way to know in advance if this will happen. Though often hailed as a precise method, the final stage of placing the new gene into a receiving higher organism is rather crude, seriously lacking both precision and predictability. The "new" gene can end up anywhere, next to any gene or even within another gene, disturbing its function or regulation. Often, genetic engineering will not only use the information of one gene and put it behind the promoter of another gene, but will also take bits and pieces from other genes and other species. Although this is aimed to benefit the expression and function of the "new" gene, it also causes more interference and enhances the risk of unpredictable effects. Consequently, the science of genetic engineering is in its infancy and has much to learn.

genome: The entire set of genetic instructions found in a cell. In humans, the genome consists of 23 pairs of chromosomes, found in the nucleus, as well as a small chromosome found in the cells' mitochondria. These chromosomes, taken together, contain approximately 3.1 billion bases of DNA sequence.

habits: Routines of behavior that are repeated regularly and tend to occur subconsciously, without directly thinking consciously about them. Habitual behavior sometimes goes unnoticed because it is often unnecessary to engage in self-analysis when undertaking routine tasks. Habituation is an extremely simple form of learning, in which an organism, after a period of exposure to a stimulus, stops responding to that stimulus in varied manners. Habits are sometimes compulsory. Behavioral habits are retained in our subconscious mind, and can be changed with mind-body interventions, such as those provided in this book.

happiness: A state of mind or feeling characterized by inner peace, love, and joy. *Natural happiness* is generated by our genetic spirituality and is independent of external circumstances such as pleasure, satisfaction, and well-being. Although a variety of philosophical, religious, psychological and biological approaches have striven to define happiness and identify its sources, the sources are always dependent on external circumstances, and, consequently, subject to change. Because natural happiness is predisposed by our spiritual genes, our spiritual expression is instantly rewarded with positive emotions. Natural happiness is self-evident by simply performing a selfless act, such as compassion or a good deed.

hypnosis: A mental state usually induced by a procedure known as a *hypnotic induction*, which is commonly composed of a series of preliminary instructions and suggestions. Hypnotic suggestions may be delivered by a hypnotist in the presence of the subject, or may be self-administered (self-hypnosis). The use of hypnotism for therapeutic purposes is referred to as *hypnotherapy*. Contrary to a popular misconception—that hypnosis is a form of unconsciousness resembling sleep—contemporary research suggests that it is actually a wakeful state of focused attention and heightened suggestibility, with diminished peripheral awareness. Self-hypnosis is used this book to help in developing happiness habits.

hypnotherapy: Often applied in order to modify a subject's behavior, emotional content, and attitudes, as well as a wide range of conditions including dysfunctional habits, anxiety, stress-related illness, pain management, and personal development. Hypnotherapy brings about deep relaxation and an altered state of consciousness, also known as a *trance*. Many people routinely experience a trance-like state while they are merely watching television or sitting at a red light. A person in a trance or deeply focused state is unusually responsive to an idea or image, but this does not mean that a hypnotist can control the person's mind and free will. On the contrary, hypnosis can actually teach people how to master their own states of awareness. By doing so they can affect their own bodily functions and psychological responses.

During hypnosis, a person's body relaxes while his or her thoughts become more focused and attentive. Similar to other relaxation techniques, hypnosis decreases blood pressure and heart rate, and alters certain types of brainwave activity. In this relaxed state, a person will feel very at ease physically yet fully awake mentally. In this state of deep concentration people are highly responsive to suggestion.

instinct: The inherent predisposition toward a particular behavior. An organism's fixed action patterns are unlearned and inherited, and the stimuli can be variable, due to imprinting in a sensitive period, or genetically fixed. Examples of instinctual fixed action patterns can be observed in the behavior of animals that perform various activities that are not based upon prior experience (such as sea turtles, hatched on a beach, that automatically move toward the ocean).

According to Darwin's Theory of Evolution, a favorable trait, such as an instinct, will be selected through competition and improved survival rate of life forms possessing the instinct. Thus, for evolutionary biology, instincts can be explained in terms of behaviors that favor survival. Human traits that have been looked at as instincts are: sleeping, altruism, disgust, face perception, "fight or flight," and competition. Some experiments in human and primate societies have also come to the conclusion that a sense of fairness could be considered instinctual (such as the way humans and apes are willing to harm their own interests in protesting unfair treatment of self or others).

meditation: A proven alternative therapy classified as a form of mind-body medicine. More and more doctors are prescribing meditation as a way to lower blood pressure, improve exercise performance in people with angina, aid breathing in people with asthma, relieve insomnia, and generally help people relax from the everyday stresses of life. Meditation is a safe and simple way to balance a person's physical, emotional, and mental states. The use of meditation for healing is not new. Meditative techniques are the product of diverse cultures and peoples around the world. Meditation is rooted in the traditions of the world's great religions and spiritual traditions, and is practiced commonly in one form or another. The value of meditation to alleviate suffering and promote healing has been known and practiced for thousands of years.

mind-body interventions: *Mind–body interventions* is the name of a U.S. National Center for Complementary and Alternative Medicine (NCCAM) classification that covers a variety of techniques designed to enhance the mind's capacity to affect bodily function and symptoms. Some of these are techniques that were once considered to be complementary or alternative medicine but have now become mainstream (for example, patient support groups and cognitive behavioral therapy). Some other mind-body techniques considered to be complementary or alternative medicine include meditation, prayer, mental healing, hypnosis, and movement re-education. Mind-body interventions are practiced based on the belief that mind, body, and spirit are connected with one another and environmental influences. Mind-body medicine aims to improve physical, mental, and emotional well-being. Stress and depression contribute to the development of and hinder recovery from chronic diseases because they create measurable hormonal imbalances. A well-known example of the mind-body interaction is the placebo effect—improved health and favorable physical changes in response to inactive medication such as a sugar pill—which confirms the connection between mind and body.

mindfulness: Calm awareness of one's body functions, feelings, content of consciousness, or consciousness itself. Mindfulness techniques are increasingly being employed in Western psychology to help alleviate a variety of mental and physical conditions. Buddhists hold that more than 2,500 years ago, Buddha provided guidance on establishing mindfulness. Mindfulness involves bringing one's awareness to focus on experience within the mind at the present moment—away from the past, the future, or some disconnected train of thought. By watching the unconscious mind, one sees that the mind is continually full of chattering, with thoughts of commentary or judgment that are connected to the past and future and therefore illusionary. As you more closely watch your mind, you will find that the cessation of suffering is not brought about by a change in outer circumstances, but rather starts with releasing attachment to thoughts, desires, predispositions, and "scripts."

mutations: Changes in the DNA sequence of a cell's genome. They are caused by environmental influences such as toxins, viruses, radiation, and genetic copy errors during DNA replication. Such mutations can prevent the gene from functioning, or change the RNA or protein produced by the gene. These alterations can have a disastrous effect, by not preventing viruses from evading the human immune system, and causing diseases.

Mutations can have a substantial effect in the process of evolution by allowing large amounts of DNA to become duplicated. These duplications are a major source of raw material for evolving novel genes, Throughout a long period of time. Most genes belong to larger families of genes of shared ancestry, but novel genes can be produced by the duplication and mutation of ancestral genes. These new combinations can enable novel traits and new functions. If these novel traits enhance survival of the species, they will in time become ancestral genes.

positive psychology: According to the Positive Psychology Center at Penn State University, positive psychology is the scientific study of the strengths and virtues that enable individuals and communities to thrive. The Positive Psychology Center promotes research, training, education, and the dissemination of positive psychology. This field is founded on the belief that people want to lead meaningful and fulfilling lives, to cultivate what is best within them, and to enhance their experiences of love, work, and play.

Positive psychology has three central concerns: positive emotions, positive individual traits, and positive institutions. Understanding positive emotions entails the study of contentment with the past, happiness in the present, and hope for the future. Understanding positive individual traits consists of the study of strengths and virtues, such as the capacity for love and work, courage, compassion, resilience, creativity, curiosity, integrity, self-knowledge, moderation, self-control, and wisdom. Understanding positive institutions entails the study of the strengths that foster better communities, such as justice, responsibility, civility, parenting, nurturance, work ethic, leadership, teamwork, purpose, and tolerance.

Author's Note: *Focusing on the positive potential of psychology, rather than traditional mental disorders, certainly seems a large advance for psychology. However, it is allied with the common misunderstanding of happiness, which always comes from without, and consequently is dependent on the variability of outside circumstances. In comparison, this book defines the state of lasting happiness as the natural result of expressing your spiritual genes, which are independent of outside influences.*

psyche: In psychoanalysis and other forms of depth psychology, *psyche* refers to the forces in an individual that influence thought, behavior, and personality, and refers to the concept of the self, encompassing the modern ideas of soul, self, and mind.

Sigmund Freud, the creator of psychoanalysis, believed that the psyche was composed of three components: the **id**, which represents the instinctual drives of an individual and remains largely unconscious; the **ego**, which is conscious and serves to integrate the drives of the id with the prohibitions of the super-ego (Freud believed this conflict to be at the heart of neurosis); and the **super-ego**, which represents a person's conscience and one's internalization of societal norms and morality.

relaxation response: A state of deep rest that reduces the physical and emotional responses of stress. It is the opposite of the fight-or-flight response. The relaxation response undoes what stress has been doing to you. It brings about decreased muscle tension, lowered heart rate and blood pressure, a deeper breathing pattern, and a peaceful, pleasant mood. The technique was developed by Herbert Benson, MD, at Harvard Medical School, tested extensively, and written up in his book entitled *The Relaxation Response*.

religion: A religion is usually equated with an organization that systemizes the expression of faith and belief in a higher power or truth. The main components of religions are beliefs, rituals, and ethics that are supposed to give meaning to the practitioner's experience of life. The term *religion* refers to both the personal practices related to communal faith and group rituals, and communication stemming from shared conviction. *Religion* is sometimes used interchangeably with *faith* or *a belief system*, but it is more socially defined than personal convictions, and it entails specific behaviors.

The development of religion has taken many forms in various cultures. It considers psychological and social roots, along with origins and historical development. Although there have been hundreds of religions throughout the ages, the more significant religions are described as a communal system for the coherence of beliefs focusing on a God, moral codes, practices, traditions, rituals, and scriptures.

Members of an organized religion may not see any significant difference between religion and spirituality, or they may see a distinction between the mundane, earthly aspects of their religion and its spiritual dimension. Some individuals draw a strong distinction between religion and spirituality; they may see spirituality as a belief in ideas of religious significance (such as God, the soul, or heaven), but not feel bound to the bureaucratic structure and creeds of a particular organized religion. They choose the term *spirituality* rather than *religion* to describe their form of belief, perhaps reflecting disillusionment with organized religion, and their movement toward a more "modern," more tolerant, and more intuitive form of religion.

self-hypnosis: A naturally occurring state of mind that can be defined as a heightened state of focused concentration (trance), with the willingness to follow instructions (suggestibility). Essentially, the method is hypnosis induction performed on you by yourself.

self-transcendence: In everyday language, *transcendence* means "going beyond," and *self-transcendence* means going beyond a prior form or state of yourself. Mystical experience is thought of as a particularly advanced state of self-transcendence, in which the sense of a separate self is abandoned. Self-transcendence is believed to be psychometrically measurable, and (at least partially) inherited. The discovery of this is described in the book *The God Gene* by Dean Hamer.

sociobiology: This term was introduced in E.O. Wilson's *Sociobiology: The New Synthesis* (1975) as the application of evolutionary theory to social behavior. Sociobiologists claim that many social behaviors have been shaped by natural selection for reproductive success, and they attempt to reconstruct the evolutionary histories of particular behaviors or behavioral strategies.

For instance, evolutionary biologists have long been puzzled by cases of apparent altruism in certain animal societies, such as sterile workers in insect colonies, warning calls that expose the caller, and food- and other resource-sharing. Such behaviors appear to be selfless and incur a cost to the actor, seemingly at odds with the self-preservational nature of natural selection. To explain this paradox, sociobiologists researched the conditions under which altruistic behavior might be consistent with the evolutionary process, and determined that, while one organism might sacrifice itself, its action benefited its family or community.

Critics of sociobiology often complain that its reliance on genetic determinism, especially of human behavior, provides tacit approval of the status quo. For example, if male aggression is genetically fixed and reproductively advantageous, then male aggression seems to be a biological reality about which we have little control.

spirituality: Relating to, consisting of, or having the nature of spirit; not tangible or material. As this definition shows, spirituality does not have a universal definition, but is dependent on your perspective. It is traditionally associated with religion, deities, the supernatural, and an afterlife. It may include existentialism and introspection, and the development of an individual's inner life through practices such as meditation, prayer, and contemplation.

Traditionally, religions have regarded spirituality as an integral aspect of religious experience, but the growth of secularism in the Western world has given rise to a broader view of spirituality. Secular spirituality carries connotations of an individual having a spiritual outlook, which is more personalized, less structured, more open to new ideas/influences, and more pluralistic than that of the doctrinal faiths of organized religions.

For a Christian, to refer to him- or herself as "more spiritual than religious" may imply relative deprecation of rules, rituals, and tradition, while preferring an intimate relationship with God. The basis for this belief is that Jesus Christ came to free humankind from those rules, rituals, and traditions, giving us the ability to "walk in the spirit," thus maintaining a "Christian" lifestyle through a one-to-one relationship with God.

Stone Age: A broad prehistoric time period during which humans widely used stone for tool-making. Stone tools were made from a variety of different sorts of stone. For example, flint was shaped (or chipped) for use as cutting tools and weapons, whereas basalt and sandstone were used for ground stone tools.

A series of metal technology innovations characterize the later Copper Age, Bronze Age, and Iron Age. The period from 2.7 to 2.58 million years ago encompasses the first use of stone tools in Gona, Ethiopia, and its spread and widespread use elsewhere soon thereafter. It ends with the development of agriculture, the domestication of certain animals, and the smelting of copper ore to produce metal. It is termed *prehistoric* because humanity had not yet started writing.

subconscious mind: The term *subconscious* is used in many different ways and has no single definition. In everyday speech and popular writing, however, the term is very commonly encountered. There it will be employed to refer to a part of the mind separate from the conscious mind. In Freud's structural theory of the psyche, the subconscious mind was termed the id, which represents the instinctual drives of an individual, and remains largely subconscious (unconscious).

At different times, references to the subconscious as an agency may credit it with various abilities and powers that exceed those possessed by consciousness; the subconscious may apparently remember, perceive, and determine things beyond the reach or control of the conscious mind. The idea of the subconscious as a powerful or potent agency has allowed the term to become prominent in the New Age and self-help literatures, in which investigating or controlling its supposed knowledge or power is seen as advantageous. The subconscious may also be supposed to contain (thanks to the influence of the psychoanalytic tradition) any number of primitive or otherwise disavowed instincts, urges, desires, and thoughts.

Bibliography

"About the Human Genome Project." *www.ornl.gov/ sci/techresources/Human_Genome/project/about.shtml.* (Accessed February 2010.

Agor, Weston. *Intuitive Management: Integrating Left and Right Brian.* Englewood Cliffs, N.J.: Alba House Cassettes, 1987.

Alleyne, Richard. "Childhood Abuse Leaves Body Physically Vulnerable to Mental Illness." *www .telegraph.co.uk.* Accessed February 22, 2009.

American Psychiatric Association. *Quick Reference to the Diagnostic Criteria From DSM-IV.* Washington, D.C.: American Psychiatric Assocation, 2000.

Armstrong, Karen. *A History of God: The 4,000 Year Quest of Judaism, Christianity and Islam.* New York: Ballantine Books, 1993.

Axelrod, Robert. *The Complexity of Cooperation: Agent-Based Models of Competition and Collaboration.* Princeton N.J.: Princeton University Press, 1997.

Baars, B. *In the Theater of Consciousness: The Workspace of the Mind*. New York: Oxford University Press, 1997.

Baird, Dr. James. *The Mindful Meals Diet*. Lincoln, Ind.: iUniverse, 2007.

Batson, C.D. *The Altruism Question*. Hillsdale, N.J.: Erlbaum, 1991.

Behe, M. *The Edge of Evolution: The Search for the Limits of Darwinism*. New York: Free Press, 2008.

Ben-Shahar, Tal, PhD. *Happier: Learn the Secrets to Daily Joy and Lasting Fulfillment*. New York: McGraw-Hill, 2007.

Blackmore, Susan. *Consciousness: An Introduction*. Oxford, N.Y.: Oxford University Press, 2005.

Borysenko, Joan, PhD. *Minding the Body, Mending the Mind*. Cambridge, Mass.: Da Capo Press, 2007.

Braden, Gregg. *Fractal Time: The Secret of 2012 and a New World Age*. Carlsbad, Calif.: Hay House, 2009.

———. Interview on *The Sixth Sense* with Dr. Laurie Nadel. *www.webtalkradio.net*. April 7, 2008.

———. *Secrets of the Lost Mode of Prayer: The Hidden Power of Beauty, Blessing, Wisdom, and Hurt*. Carlsbad, Calif.: Hay House, 2006.

———. *The Spontaneous Healing of Belief: Shattering the Paradigm of False Limits*. Carlsbad, Calif.: Hay House, 2008.

Brigham, Deirdre Davis, with Adelaide Davis and Derry Cameron-Sampey. *Imagery for Getting Well: Clinical Applications of Behavioral Medicine*. New York: W.W. Norton and Company, 1994.

Bruce, Eve, MD. *Shaman, M.D.: A Plastic Surgeon's Remarkable Journey Into the World of Shapeshifting.* Rochester, Vt.: Destiny Books, 2002.

Buckingham, Marcus. "What's Happening to Women's Happiness?" *The Huffington Post. www.huffingtonpost.com/marcus-buckingham.* September 27, 2009.

Buss, David, Ed. *The Handbook of Evolutionary Psychology.* Hoboken, N.J.: John Wiley & Sons, 2005.

Campbell, Joseph. *The Hero With a Thousand Faces.* Princeton, N.J.: Pantheon Books, 1949.

Carey, G. *Human Genetics for the Social Sciences.* Thousand Oaks, Calif.: Sage Publications, 2003.

Carson, Ronald A., Mark A. Rothstein, and Floyd E. Bloom. *Behavioral Genetics: The Clash of Culture and Biology.* Baltimore, Md.: The Johns Hopkins University Press, 2002.

Chalmers, David. *The Conscious Mind: In Search of a Fundamental Theory.* New York: Oxford University Press, 1996.

Church, Dawson, PhD. *Epigenetic Medicine and the New Biology of Intention, 2nd Edition.* Santa Clara, Calif.: Elite Books, 2009.

Clark, W., and M. Grunstein. *Are We Hardwired? The Role of Genes in Human Behavior.* New York: Oxford University Press, 2000.

Cleermans, A., ed. *The Unity of Consciousness: Binding, Integration, and Dissociation.* New York: Oxford University Press, 2000.

Cobb, John. *The Structure of Christian Existence.* Lanham, Md.: University Press of America, 1967.

Collins, F. *The Language of God*. New York: Free Press, 2006.

Crews, David. Definition from "Epigenetics and its implication for behavioral neuroendocrinology." *Frontiers in Neuroendocrinology* 29, No. 3: 334–57 (2008). *http://cat.inst.fr/aModele=afficheN@cpsidt=20391246*.

Crockett, Tom. *Stone Age Wisdom: The Healing Principles of Shamanism*. Gloucester, Mass.: Fair Winds Press, 2003.

Czikszentmihalyi, Mihaly. "Finding Flow." *Psychology Today*. July–August 1997, p. 46.

Dawkins, R. *The Blind Watchmaker: Why the Evidence from Evolution Reveals a Universe Without Design*. New York: W.W. Norton, 1987 and 1996.

Dawkins, Richard. *The Selfish Gene*. New York: Oxford University Press, 1976 and 1989.

De Beauport, Elaine, with Aura Sofia Diaz. *The Three Faces of Mind: Developing Your Mental, Emotional, and Behavioral Intelligences*. Wheaton, Ill.: Quest Books/Theosophical Publishing House, 1996.

Dembski, W. *No Free Lunch: Why Specified Complexity Cannot Be Purchased Without Intelligence*. Lanham, Md.: Rowman & Littlefield Publishers, 2007.

Dick, D.M., and R.J. Rose. "Behavior genetics: What's new? What's next." *Current Directions in Psychological Science* 11, No. 2: 70–4 (2002).

Diener, Ed. *The Science of Well-Being: The Collected Works of Ed Diener*. New York: Springer, 2009.

Diener, Ed, and Robert Biswas-Diener. *Happiness: Unlocking the Mysteries of Psychological Wealth*. Malden, Mass.: Blackwell Publishing, 2008.

DNA Initiative: Advancing Criminal Justice Through DNA Technology Website. *www.dna.gov*.

Donaldson, Dwight M. *Studies in Muslim Ethics*. London: S.P.C.K, 1963, p. 82.

Dossey, Larry, MD. *The Power of Premonitions: How Knowing the Future Can Shape Our Lives*. New York: Dutton/Penguin, 2009.

Doyle, Alister. "Botswana 'snake rock' may show Stone Age religion." Oslo: Reuters News Service, November 30, 2008.

Dunbar, Robin. *The Human Story*. London: Faber and Faber, 2005.

Dyer, Dr. Wayne W. *Change Your Thoughts, Change Your Life: Living the Wisdom of the Tao*. Carlsbad, Calif.: Hay House, 2007.

"Epigenetics means that what we eat, how we live and love alters how our genes behave." Duke Medicine News and Communications press release, October 25, 2005.

"Eugenics." Dictionary.com. *http://dictionary.reference.com/browse/eugenics*. Accessed February 2010.

Evans, Dylan, and Oscar Zarate. *Introducing Evolutionary Psychology, 2nd Edition*. Cambridge, UK: Totem Books, 2005.

Evans, J.D, and D.E. Wheeler. "Differential gene expression between developing queens and workers in the honey bee, apismellifera." Proceedings of the National Academy of Sciences 96: 5575–80 (1999).

Fagg, L. *Electromagnetism and the Sacret: At the Frontier of Spirit and Matter*. New York: Continuum, 1999.

Frankl, Victor E. *Man's Search for Meaning: An Introduction to Logotherapy, 4th Edition*. Boston: Mass.: Beacon Press, 1992.

Gallup, George Jr. *Religion in America*. Princeton, N.J.: Princeton Religion Research Center, 1985.

Gallup, Jr. George, and Jim Castelli. *The People's Religion: American Faith in the '90s*. New York: Oxford University Press, 1996.

Gardner, Herb. *A Thousand Clowns: A Comedy in Three Acts*. New York: Samuel French, Inc., 1961.

Ghiselin, Michael T. "Darwin and Evolutionary Psychology." *Science* 179: 964–8 (1973).

Gilbert, Daniel. *Stumbling on Happiness*. New York: Vintage Books, 2007.

Goleman, D. *Social Intelligence: The New Science of Human Relationship*. New York: Bantam, 2006.

Gonzales, G., and J. Richards. *The Privileged Planet: How Our Place in the Cosmos Is Designed for Discovery*. Washington, D.C.: Regnery Publishing, Inc., 2004.

Gottesman, I., S. Brenner, and J.H. Miller, eds. "Nature-nurture controversy." *Encyclopedia of Genetics* 3: 1297–1302 (2002). London: Academic Press.

Greeley, Andrew M. *Religion in Britain, Ireland and the USA*. New York: Guilford Press, 2007.

Grof, Stanislav. *Primal Renaissance: The Journal of Primal Psychology* 2, No. 1 (Spring 1996).

Grove, Hannah Saw, and Russ Alan Prince. *Facts & Figures: Money as Aphrodisiac—Being Rich Means Getting Lucky on Your Terms*. Edition 1, January 2007.

Guralnik, David, ed. *Webster's New World Dictionary of the American Language*. New York: World Publishing Company, 1975.

Hagerty, Barbara Bradley. *Fingerprints of God: The Search for the Science of Spirituality*. New York: Riverhead Books, 2009.

Hamer, Dean. *The God Gene: How Faith Is Hardwired Into Genes*. New York: Doubleday, 2004.

Hamer, Dean H., and Peter Copeland. *Living With Our Genes: Why They Matter More Than You Think*. New York: Random House, 1999.

"Happiness." Wikipedia.org. *http://en.wikipedia.org/wiki/ Happiness*. Accessed February 2010.

Harris, A.H, F.M. Lufkin, S.V. Benisovich, S. Standard, J. Bruning, S. Evans, and C. Thoresen. "Effects of a group forgiveness intervention on forgiveness, perceived stress and trait anger: A randomized trial." *Journal of Clinical Psychology* 62, No 6: 715–33 (2006).

Herrigel, Eugen. *Zen in the Art of Archery*. New York: Pantheon Books, 1953.

Holden, Robert, PhD. *Be Happy: Release the Power of Happiness in You*. Carlsbad, Calif.: Hay House, 2009.

Huffington, Arianna. *The Fourth Instinct: The Call of the Soul*. New York: Simon and Schuster, 1994.

Ilibagiza, Immaculee. *Left to Tell: Discovering God Amidst the Rwandan Holocaust*. Carlsbad, Calif.: Hay House, 2006.

Jablonka, Eva, and Marion J. Lamb. *Evolution in Four Dimensions: Genetic, Epigenetic, Behavioral, and Symbolic Variation in the History of Life*. Cambridge, Mass.: The MIT Press, 2005.

Jaynes, Julian. *The Origin of Consciousness in the Breakdown of the Bicameral Mind*. New York: Mariner Books, 2000.

Kant, Immanuel. *Critique of Pure Reason*. Amherst, N.Y.: Prometheus Books, 1990.

Kent, James. "The Cure for Depression? Have Fun!" *Mind Power News*, October 26, 2009. *www .mindpowernews.com/FunDeficiency.htm*.

Klein, Stefan, and Stephen Lehmann. *The Science of Happiness: How our Brains Make us Happy, and What We Can Do to Get Happier*. New York: Marlowe & Company, 2006.

Kluger, Jeffrey. "Is God in Our Genes?" *Time*, October 25, 2004, p. 66.

Koch, C. *The Quest for Consciousness*. Englewood, Colo.: Roberts & Company, 2004.

Kornfield, Jack. *The Roots of Buddhist Psychology*. Boulder, Colo.: Sounds True, 1999.

Kosmin, Barry S., and Seymour P. Lachman. *One Nation Under God: Religion in Contemporary American Society*. New York: Crown, 1993.

Lama, Dalai. *The Universe in a Single Atom*. New York: Morgan Road Books, 2005.

Layard, R. *Happiness Lessons From a New Science*. New York: Penguin Press, 2005.

Leonhardt, D. "If Richer Isn't Happier, What Is?" *New York Times*, May 19, 2001, B9–11.

Libet, Benjamin, Anthony Freeman, and Keith Sutherland, eds. *The Volitional Brain: Towards a Neuroscience of Free Will*. Exeter, UK: Imprint Academic, 1999.

Lipton, Bruce H., PhD. *The Biology of Belief: Unleashing the Power of Consciousness, Matter, & Miracles*. Carlsbad, Calif.: Hay House, 2008.

———. *Spontaneous Evolution: Our Positive Future (And a Way to Get There From Here)*. Carlsbad, Calif.: Hay House, 2009.

Livescience staff. "Happiness May Be Inherited." *LiveScience.com*, May 24, 2009. *www.livescience.com/health/090514-happy-inheritance.html*.

"Map of World Happiness: A Global Projection of Subjective Well-Being." Technovelgy.com. *www.technovelgy.com/ct/Science-Fiction-News.asp?NewsNum=893*. Accessed February 2010.

McCraty, Rollins. Interview with Dr. Laurie Nadel, August 7, 2009.

McTaggart, L. *The Field: The Quest for the Secret of the Universe*. New York: HarperCollins, 2008.

Metzinger, T. *Being No One: The Self-Model Theory of Subjectivity*. Cambridge, Mass.: MIT Press, 2003.

Metzinger, T., ed. *The Neural Correlates of Consciousness*. Cambridge, Mass.: MIT Press, 2000.

Mill, John. *Comte, August, Catechism and Positivism*. London: Tutis Digital Publishing, 2007.

Miller, G.E. "Can Money Buy Happiness?" *Twenty-Something Finance*. January 19, 2009. *http://20somethingfinance.com/blog/2009/01/19/can-money-buy-happiness-5-reasons-why- it-could-but-may-not/*.

Mishlove, Jeffrey, PhD. *The Roots of Consciousness: The Classic Encyclopedia of Consciousness Studies, Revised and Expanded*. Tulsa, Okla.: Council Oak Books, 1993.

Mitchell, Dr. Edgar. Interview on *The Sixth Sense* with Dr. Laurie Nadel. *www.webtalkradio.net*, June 22, 2008.

Mitchell, Dr. Edgar, with Dwight Williams. *The Way of the Explorer: An Apollo Astronaut's Journey Through the Material and Mystical Worlds, Revised Edition.* Franklin Lakes, N.J.: Career Press/New Page Books, 2008.

Mitroff, Ian. *Smart Thinking for Crazy Times: The Art of Solving the Right Problems.* San Francisco: Berret-Koehler Publishers, Inc, 1998.

"Mood May Be Written In (and On) Genes." ABC News. *http://abcnews.go.com/pring?id=4839669* Accessed October 2009.

Moreira-Almeida, Alexander, Francisco Lotufo Neto, and Harold G. Koenig. "Religiousness and mental health: a review." *Rev. Bras. Psiquiatr* 28, No. 3: 242–50 (September 2006). *www.scielo.br/scielo.php?script=sci_arttext&pid=S1516-44462006000300018&lng=en&nrm=iso.*

Myss, Caroline. *Defy Gravity: Healing Beyond the Bounds of Reason.* Carlsbad, Calif.: Hay House, 2009.

Nadel, Laurie, PhD. "Make Way For Joy: Little Things You Can Do to Find Peace in Your Life." *Woman's Day.* June 1, 2007.

Nadel, Laurie, PhD, with Judy Haims and Robert Stempson. *Dr. Laurie Nadel's Sixth Sense: Unlocking Your Ultimate Mind Power.* Lincoln, Neb.: ASJAPress/iUniverse, 2007.

National Academy of Sciences. *Science and Creationism: A View From the National Academy of Sciences, 2nd Edition*. Washington, D.C.: National Academy Press, 1999.

"National Accounts of Well-Being." NEF Website. *www.nationalaccountsofwellbeing.org*.

Newberg, Andrew, Eugene D'Aquili, and Vince Rause. *Why God Won't Go Away: Brain Science and the Biology of Belief*. New York: Ballintine, 2002.

Niemiec, Christopher, Richard Ryan, and Edward Deci. "The path taken: Consequences of attaining intrinsic and extrinsic aspirations in post-college life." *Journal of Research in Personality* 43, Issue 3: 291–306 (June 2009).

Nixon, Robin. "Epigenetics: A revolutionary look at how humans work." *LiveScience.com* April 27, 2009. *www.livescience.com/health/090427-overview.html*.

Oliner, Samuel P., and Pearl M. Oliner. *Towards a Caring Society: Ideas into Action*. West Port, Conn.: Praeger, 1995.

Oord, Thomas Jay. *The Altruism Reader: Selections from Writings on Love, Religion, and Science*. Philadelphia: Templeton Foundation Press, 2007.

———. *Science of Love: The Wisdom of Well-Being*. Philadelphia: Templeton Foundation Press, 2004.

Ornish, D., et al. "Changes in prostate gene expression in men undergoing an intensive nutrition and lifestyle intervention." Proceedings of the National Academy of Sciences 17:105 (24), p. 8369 (2008 June).

Osho Active Meditation Series. *Joy, The Happiness That Comes from Within*. New York: St. Martin's Press, 2004.

Parens, Erik, Audrey Chapman, and Nancy Press. *Wrestling With Behavioral Genetics: Science, Ethics, and Public Conversation*. Baltimore: The John Hopkins University Press, 2006.

Pattakos, Alex, PhD. *Prisoners of Our Thoughts: Viktor Frankl's Principles for Discovering Meaning in Life and Work, 2nd Edition*. San Francisco: Berret-Koehler Publishers, Inc., 2008. *www.prisonersofourthoughts.com*.

Pert, Candace B., PhD. *Molecules of Emotion: The Science Behind Mind-Body Medicine*. New York: Simon and Schuster, 1999.

Pert, Candace B., PhD, with Nancy Marriott. *Everything You Need to Know to Feel Go(o)d*. Carlsbad, Calf.: Hay House, 2006.

Peters T., and M. Hewlett. *Evolution From Creation to New Creation: Conflict, Conversation, and Convergence*. Nashville, Tenn.: Abingdon Press, 2003.

Plomin, R., J.C. DeFries, G.E. McClearn, and M. Rutter. *Behavioral Genetics, 3rd Edition*. New York: Worth Publishers, 2008.

Plomin, Robert, John C. Defries, Ian W. Craig, and Peter McGuffin. *Behavioral Genetics in the Postgenomic Era*. Washington, D.C.: American Psychological Association, 2002.

Puchalski, Christina M. "Spirituality, Religious Wisdom, and the Care of the Patient Forgiveness: Spiritual and Medical Implications." *The Yale Journal for Humanities in Medicine*. September 17, 2002.

Rachels, James, and Stuart Rachels. *The Elements of Moral Philosophy*. New York: McGraw-Hill, 2009.

Rand, Ayn. *The Virtue of Selfishness*. New York: Signet, 1964.

Reagan, M. *The Hand of God: Thoughts and Images Reflecting the Spirit of the Universe*. Atlanta, Ga.: Lionheart Books, 1999.

Rosenstand, Nina. *In The Moral of the Story*. Mountain View, Calif.: Mayfield Publishing, 2000.

Rost, H.T.D. *The Golden Rule: A Universal Ethic*. Oxford, UK: George Ronald, 1986.

Salk, Jonas. *Anatomy of Reality: Merging of Intuition and Reason*. New York: Columbia University Press, 1983.

Schaafsma, P. *Indian Rock Art of the Southwest*. Albuquerque, N.M.: New Mexico School of American Research and the University of New Mexico Press, 1980.

Schroeder G. *The Science of God: The Convergence of Scientific and Biblical Wisdom*. New York: Free Press, 1997.

Schwartz, Gary, and William L. Simon. *The G.O.D. Experiments: How Science is Discovering God in Everything, Including Us*. Philadelphia: Templeton Foundation Press, 2007.

Seligman, Martin. *The Optimistic Child*. New York: Houghton Mifflin, 1996.

Seligman, Martin E.P. *Authentic Happiness: Using the New Positive Psychology to Realize Your Potential for Lasting Fulfillment*. New York: Free Press, 2002.

Shermer, M. *How We Believe: Science, Skepticism and the Search for God*. New York: Henry Holt and Company, 2000.

Smedes, Lewis B. *Forgive and Forget: Healing the Hurts We Don't Deserve*. New York: HarperCollins, 1984.

Smith, H. *Why Religion Matters: The Fate of the Human Spirit in an Age of Disbelief*. New York: HarperCollins, 2001.

Spetner, L. *Not by Chance: Shattering the Modern Theory of Evolution*. Brooklyn, N.Y.: Judaica Press, 1997.

"Spirituality." Dictionary.com. *http://dictionary.reference.com/browse/spirituality*.

Sutherland, K. "Spirits From the South." *The Artifact* 34, Nos. 1&2 (1996). El Paso Archaeological Society.

"Symptoms of major depressive disorder." *American Psychiatric Association Diagnostic Manual (DSM IV), Fourth Edition*. Washington, D.C.: American Psychiatric Association, 2004.

Tost, J. *Epigenetics*. Norwich, UK: Caister Academic Press, 2008.

Turkheimer, E. "Three laws of behavior genetics and what they mean." *Current Directions in Psychological Science* 9: 160–4 (2000).

Turner, Bryan. *Chromatin and Gene Regulation: Mechanisms in Epigenetics*. Malden, Mass.: Blackwell Publishing, 2002.

Veenhoven, R. *Bibliography of Happiness*. Rotterdam, Netherlands: Erasmus University, 1993.

———. "Happiness in nations: subjective appreciation of life in 56 nations, 1946–1992." *RISBO*. Erasmus University, Rotterdam (1993) ISBN 90-72597-46-X, 365 pages.

———. "World Database of Happiness." *Distributional Findings in Nations*, Erasmus University, Rotterdam. *http://worlddatabaseofhappiness.eur.nl* (2009).

Ward, Keith. *Is Religion Dangerous?* Oxford, UK: Lion Hudson, 2006.

Watson, James D. *Avoid Boring People: Lessons From a Life in Science*. New York: Alfred E. Knopf, 2007.

Wattles, Jeffrey. *The Golden Rule*. New York: Oxford University Press, 1996.

Welch, John, and O. Carm. *The Mystics and the Development of Consciousness*. Canfield, Ohio: Alba House Cassettes, 1987.

"What Is Epigenetics?" *http://epigenome.eu/en/1,1,0*.

"Who's Happy and Why." Mayo Clinic press release. December 9, 2008. *www.mayoclinic.org/news2008-mchi/5111.html*.

Whybrow, Peter. *American Mania: When More Is Not Enough*. New York: W. Norton, 2005.

Wilkinson, Will. *In Pursuit of Happiness Research: Is it Reliable? What Does it Imply for Policy?* Policy Analysis No. 590. The Cato Institute, 2007.

Willis, Gary. *Under God: Religion and American Politics*. New York: Simon and Schuster, 1990.

Wilson, Edward O. *Consilience: The Unity of Knowledge*. Cambridge, Mass.: Harvard University Press, 1998.

———. *On Human Nature*. Cambridge, Mass.: Harvard University Press, 1978.

———. *Sociobiology: The New Synthesis, 25th Anniversary Edition*. Cambridge, Mass.: Harvard University Press, 2000.

"The World Map of Happiness." *http://blog .guykawasaki.com/2007/02/the_world_map_o .html#ixzz0SuHD4U9a*. (The University of Leicester site with original data is restricted by password to university students and professors.)

Wright, Robert. *The Moral Animal: Why We Are the Way We Are; The New Science of Evolutionary Psychology*. New York: First Vintage Books, 1995.

Zimmerman, Bill. *Make Beliefs*. New York: Guarionex Press, 1987.

Internet Resource Directory

For more information about epigenetics, happiness research, and the experts interviewed in this book, explore the following Internet resources.

www.barbarabradleyhagerty.com: A religion correspondent for National Public Radio and the author of *Fingerprints of God: The Search for Science in Spirituality*, Barbara Bradley Hagerty's Website has book reviews, audio interviews, and news of her schedule.

http://blog.guykawasaki.com/2007/02/the_world_map_0 .html#ixzz0SuHD4U9a: Summary of the World Map of Happiness. (The University of Leicester site with original data is restricted by password to university students and professors.)

www.brucelipton.com: Margaret Horton and Sally Evans maintain this site to keep up with Bruce's travel schedule, new products, and studies. Keeping track of Bruce is a full-time job!

www.cshl.org: Cold Spring Harbor Laboratory official site.

www.dna.gov: Everything you need to know about DNA.

www.epigenome.eu: A well-organized European Website with background information and a chronology of research in epigenetics.

www.genieinyourgenes.com: News about Dawson Church, PhD's best-selling book.

www.genomics.energy.gov: Continually updated with new findings on the Human Genome Project.

www.globalcoherence.org: Global Coherence Initiative project is the not-for-profit branch of Heart Math Institute. Here you will find experiments that you can do at home and updated studies on emWave programs and global coherence.

www.greggbraden.com: Information on Gregg Braden's latest books, workshops, and public appearances, with links to HeartMath.org.

www.happinessgenes.org: A resource site for free reports and information; authors' contact information and whereabouts.

www.heartmath.org: A comprehensive Website with updated studies on emWaves, heart rate variability and DNA, health issues, stress, and intuition.

www.mayoclinic.org: Check this Website for studies on wellness/stress, genetics, and happiness.

www.mindpowernews.com: A weekly e-zine digest of articles on scientific research on the mind, spirituality, remote viewing, and future developments.

www.nationalaccountsofwellbeing.com: Here you can fill out a survey and receive a spider web-shaped graph that shows how you score on 10 components of well-being and happiness. You can superimpose your profile on a composite European profile and/or profiles from European countries.

www.neweconomics.org: A "think-and-do" tank in London, the NEF (National Economic Foundation) conducts pioneering research on well-being throughout Europe.

www.prisonersofourthoughts.com: This site continually updates the work of "Dr. Meaning": Alex Pattakos, PhD's workshops and publications around the world.

www.realityshifters.org: Physicist and professional intuitive Cynthia Larson's site contains information about her work and her signature question: "How *good* can it get?"

www.shamanportal.org: From the Stone Age to the 21st century, this is an exciting site with videos of indigenous shamans performing rituals, with articles, book reviews, and products as well.

www.ssp.org: The Society of Shamanic Practitioners is a not-for-profit organization of health professionals in the United States who incorporate shamanic principles into their work.

www.spontaneousevolution.com: For news about Dr. Bruce Lipton and Steve Bhaerman, authors of *Spontaneous Evolution*, go to this site.

www.stevebhaerman.com: When he is not writing and lecturing about politics, Steve's stand-up comedy persona works the stage. If you have not yet met Swami Beyondananda, you won't be able to stop laughing.

www.stressproject.org: Dr. Church's Website contains a video on his work using the Emotional Freedom Technique (EFT) with American servicemen and women struggling with PTSD; scheduled workshops; volunteers (including Laurie Nadel, PhD) who do pro-bono sessions with veterans and people in the military.

http://worlddatabaseofhappiness.eur.nl: World Database of Happiness, Distributional Findings in Nations, compiled by R. Veenhoven at Erasmus University, Rotterdam.

Index

About the Authors

James D. Baird, PhD, has more than 40 years of experience as a successful inventor and graduate engineer. His inventor background instilled in him a persistent curiosity about how things work, translating from mechanisms to life. His passion for understanding the bioengineering that makes us human, combined with his religious and spiritual beliefs, has led him to research the subject of happiness for more than 20 years, and, in the process, earned him a PhD in Natural Health.

He was the founder of "Happiness Ministries," a not-for-profit ministry that produced inspirational texts. His first two books, *The Happiness Plan* (Liquori, 1990) and *The Modern Christian's Happiness Plan* (Wine Press, 1999) are based on the belief that natural happiness results from a belief in a supernatural power and altruistic ethics, such as agape love and compassion. Baird has been involved with many religious organizations, including nursing homes, churches, and charitable organizations. Dr. Baird is chairman of a Chicago chapter of the American Scientific Association, a national professional association whose mission is to bring harmony to science and religion.

As a natural health advocate with an inventor's curiosity, he was intrigued by the paradox that "diets don't work." In an effort to uncover the reasons and provide an answer, he wrote his third book, *The Mindful Meals Diet* (iUniverse, 2007), which explained how genetic factors are the basis of unhealthy eating habits that lead people to become overweight. His program incorporated self-hypnosis and mind/body strategies to develop healthy eating habits (*www.mindfulmealsdiet.com*).

Excited by the findings of the Human Genome Project, he had the intuition that spiritualism had a genetic basis, and that natural happiness was a design of our creator. The convincing evidence he uncovered that we are endowed with spiritual genes that motivate and emotionally reward a belief in a supernatural power and altruistic ethics was overwhelming. As to why there is still violence in the world, he concluded that it has to do with ancient instincts and subconscious thinking, which now can be treated by a new science termed *epigenetics*.

Happiness Genes: Unlock the Potential Hidden in Your DNA is his fourth book. It presents convincing evidence of our genetic spirituality, and the nature of its rewarding happiness. To help the reader develop natural happiness habits, it provides epigenetic methods that suppress the subconscious thinking that hinders spiritual expression.

Laurie Nadel, PhD, spent 20 years as a journalist for major American news organizations, including CBS News and The *New York Times* where she wrote a religion column, "Long Island at Worship." The author of the best-seller *Sixth Sense: Unlocking Your Ultimate Mind Power* (ASJA Press, 2007), she has appeared on *Oprah*. "The Dr. Laurie Show" on Genesis Communications Network explores New Science topics. (*www.gcnlive.programs/laurie*)

She holds doctorates in psychology and clinical hypnotherapy with a specialty in stress/wellness issues and Post-Traumatic Stress Disorder. (*www.laurienadel.com*)

Happiness Genes: Unlock the Potential Hidden in Your DNA is her sixth book.